Electrochemical Impedance Spectroscopy

Edited by Marwa El-Azazy, Mart Min and Paul Annus

Published in London, United Kingdom

IntechOpen

Supporting open minds since 2005

Electrochemical Impedance Spectroscopy
http://dx.doi.org/10.5772/intechopen.87884
Edited by Marwa El-Azazy, Mart Min and Paul Annus

Contributors
Haval Yacoob Aldosky, Dindar S. Bari, Jacques Sawadogo, Jean Boukari Legma, Elmelouky Abderrahmane, Salah Kouass, Hector Herrera Hernandez, Jorge Morales-Hernandez, Carlos O. Gonzalez Moran, Adriana Mercedes Ruiz Reynoso, Araceli Mandujano Ruiz, Khuloud Almzarzie, Ahmad Falah, Ayman Almassri, Hassan Kellawi, Alexandru Gabriel Gheorghe, Florin Constantinescu, Miruna Nitescu, Mihai Eugen Marin, Marwa S. El-Azazy, Juan C. Trinidad González, José G. Miranda Hernández, Ricardo Orozco Cruz, Amor Fadhalaoui, Hassouna Dhaouadi, Fathi Touat, Abdelhadi Mortadi, Elghaouti Chahid, Reddad Elmoznine

First published in London, United Kingdom, 2021 by IntechOpen
IntechOpen is the global imprint of INTECHOPEN LIMITED, registered in England and Wales, registration number: 11086078, 5 Princes Gate Court, London, SW7 2QJ, United Kingdom
Printed in Croatia

British Library Cataloguing-in-Publication Data
A catalogue record for this book is available from the British Library

Additional hard and PDF copies can be obtained from orders@intechopen.com

Electrochemical Impedance Spectroscopy
Edited by Marwa El-Azazy, Mart Min and Paul Annus
p. cm.
Print ISBN 978-1-78985-215-8
Online ISBN 978-1-78985-216-5
eBook (PDF) ISBN 978-1-78985-677-4

For EU product safety concerns:
IN TECH d.o.o., Prolaz Marije Krucifikse Kozulić 3, 51000 Rijeka, Croatia, info@intechopen.com or visit our website at intechopen.com.

Meet the editor

Dr. Marwa El-Azazy is an analytical chemist and experienced educator and researcher with more than 15 years of teaching experience at several institutions. Her main research interest is construction of sensors (using microfluidic platforms for point-of-care-testing of drugs and bio-analytes and ion-selective electrodes), chemometrics and analytical method development, spectroscopic analyses of drugs and pharmaceuticals, synthesis and characterization of nanomaterials, and development of green chemistry approaches for wastewater treatment. Dr. Marwa has a track record of research inputs including more than thirty refereed papers in prestigious international journals, several conference presentations, and two book chapters, in addition to several research grants. She serves as a reviewer for a variety of international journals.

Dr. Mart Min has been a professor and leading scientist at Thomas Johann Seebeck Department of Electronics, Tallinn University of Technology, Estonia, since 1992. He has been professor emeritus since 2017. He obtained a PhD in Measurement Science from Kiev Polytechnic, Ukraine, in 1984. From 1992 to 1993, he was with the Technical University and Bundeswehr University in Munich, Germany. During 2007–2010, Dr. Min joined the Institute of Bioprocessing and Analytical Measurement Technique, Germany. His interests include the methods for measurement of electrical impedance and processing of electrical signals. Dr. Min is a Senior Life Member of the IEEE Instrumentation and Measurement and Engineering in Medicine and Biology Societies. He is also a member of the International Committee for Promotion of Research in Bio-Impedance (ICPRBI).

Paul Annus, PhD, has been a teacher and entrepreneur since graduating in 1987. He has spent the last two decades working in engineering and project management in the local technology development center and research in the Tallinn University of Technology, where he currently works as a senior researcher, lecturer, and supervisor in the Thomas Johann Seebeck Department of Electronics. He is a member of the Institute of Electrical and Electronics Engineers. He has received awards for scientific papers, reviewing those papers, project leadership, and course development. His research interests include data acquisition, system identification, and especially bioimpedance spectroscopy. His emphasis is on reliable, scalable, and optimally deployable solutions for health monitoring. He has published more than 100 publications and registered thirteen inventions.

Contents

Preface

Electrochemical impedance spectroscopy (EIS) is a potent electrochemical technique with a variety of applications. EIS measurements involve the application of an alternating current (AC) to the system under investigation, followed by measurement of the response in the form of AC current (or voltage) as a function of frequency. In general, three factors make EIS exceptionally attractive in terms of applications: (1) capability to explore electrochemical systems (ES) at relatively low frequencies, in contrast to the majority of electrochemical techniques, which involve an application of large perturbation for sensing the membrane/electrolyte interface; (2) feasibility of implementation of EIS into the system to be measured; and (3) usefulness of data obtained in characterizing the studied ES, where EIS provides on-site information on the relaxation data over a range of frequencies, from as low as 10-4 Hz up to 106 Hz. A combination of these advantages led to the widespread use of EIS in a variety of applications.

This book introduces readers to the basic principles of EIS with a focus on recent developments and applications. The book is divided into eight chapters, which in turn coalesce the application of EIS in different fields. Chapter 1 by Hernández et al. provides readers with comprehensive information on the use of EIS in monitoring and hence controlling the corrosion of steel. In Chapter 2, Sawadogo et al. investigates the application of EIS in studying and managing the corrosion behavior of recycled aluminum alloy in various culinary media. Chapter 3 by Almzarzie et al. studies the use of a non-toxic and ecofriendly corrosion inhibitor extracted from turmeric roots. Kouass et al., in Chapter 4, use EIS to analyze the design of multi-ionic doped $CePO_4$ materials with improved ionic conductivity and electrochemical stability. Chapter 5 by Gheorghe et al. discusses the circuit models of bioelectric impedance. This chapter shows an interesting application of EIS in medicine, where in various applications accurate information about fluid distribution in different compartments of the body may lead to substantial conclusions. Chapter 6 by Aldosky et al. reviews another application of EIS as a non-invasive technique, as well as the use of electrodermal activity (EDA) as a sensitive measure of sympathetic nervous system activity. Elmelouky et al., in Chapter 7, study environmental applications of EIS and the adsorption behavior of nitrate ions by layered double hydroxide. Finally, in Chapter 8, El-Azazy conducts a thorough review of applications of EIS in food, water, and drug analysis.

I would like to use this opportunity to thank my co-editors Prof. Mart Min and Dr. Paul Annus for their great efforts in reviewing and editing the book. I am entirely indebted to all the authors for their valuable contributions. The contributors are experts in their respective fields and have tried their best to disclose the inherent concepts related to the applications covered in this volume. Their efforts to make this book comprehensive and informative are much appreciated.

I am most grateful to Author Service Manager Mr. Mateo Pulko for all his efforts and support throughout the process of editing this book. Many thanks to the entire

IntechOpen publishing team for making this book project possible. Last, but by no means least, I am grateful to my family for all the love, incessant encouragement, and support.

We hope readers will find this book informative and that it will spur further advances in the field.

Dr. Marwa El-Azazy
Department of Chemistry and Earth Sciences,
College of Arts and Sciences,
Qatar University,
Doha, Qatar

Mart Min and Paul Annus
Tallinn University of Technology,
Estonia

Chapter 1

Electrochemical Impedance Spectroscopy (EIS): A Review Study of Basic Aspects of the Corrosion Mechanism Applied to Steels

Héctor Herrera Hernández, Adriana M. Ruiz Reynoso, Juan C. Trinidad González, Carlos O. González Morán, José G. Miranda Hernández, Araceli Mandujano Ruiz, Jorge Morales Hernández and Ricardo Orozco Cruz

Abstract

AC impedance measurements have been applied for over twenty years in electrochemistry and physics to investigate the electrical properties of conductive materials and their interfaces using an external electrical impulse (*VOLTAGE, V* or *CURRENT, I*) as driving force. Furthermore, its application has recently appeared to be destined in the Biotechnology field as an effective tool for rapid microbiologic diagnosis of living organism in situ. However, there is no doubt that the electrochemical impedance spectroscopy (EIS) is still one of the most useful techniques around the world for metal corrosion control and its monitoring. Corrosion has long been recognized as one of the most expensive stumbling blocks that concern many industries and government agencies, because it is a steel destructive phenomenon that occurs due to the chemical interaction with aqueous environments and takes place at the interface between metal and electrolyte producing an electrical charge transfer or ion diffusion process. Consequently, it is experimentally possible to determine through the EIS technique the mechanism and control that kinectics of corrosion reactions encounter. First, EIS data is collected through a potentiostat/galvanostat apparatus. After, it is fitted to a mathematical model (*i.e.* an equivalent electrical circuit, EEC) for its interpretation and analysis, fundamentally seeking a meaningful physical interpretation. Finally, this review reports some basic aspects of the corrosion mechanism applied to steels through the experimental EIS response using Nyquist or Bode plots. Examples are given for different applied electrochemical impedance cases in which steel is under study intentionally exposed to a corrosive aqueous solution by applying a sinusoidal potential at various test conditions.

Keywords: electrochemical impedance spectroscopy, corrosion process, charge transfer, electrical impulse, equivalent circuit, steels, chemical reactions

1. Introduction

Corrosion of steels represents worldwide, one of the most costly problems that several industries are challenged every day due to the aggressive conditions during the manufacturing process of the steel parts or the premature failure of steel tools by stress corrosion cracking (SCC) as well as deterioration of steel components from equipment and machinery in a certain service. The construction industry is an example in where steel is essential, which requires durable and strong structures for the build of bridges, tunnels, towers, buildings, airports, roads, plants and railways. Many of these constructions are usually outdoors, exposed to the atmosphere conditions, additionally, the surrounding environment where these steels are placed for their service is often highly polluted, that it often degrades the steel structure at a considerable corrosion rate. Some of those steels are also design to be used in the; mining industry, pipeline transport of fluids, shipbuilding, agriculture equipment and heavy machinery, among others. During their usage, steels are also severely damaged by one type of corrosion mechanism [1–4]. According to Zaki Ahmad [5] the concept of corrosion must be defined taking into account the environment in which the metal-materials are place to serve for long periods of exposure time, thus, all the environments are considered corrosive to some degree of damage as follows; i) air humidity, ii) fresh, distilled, salt and marine water, iii) natural urban, marine and industrial atmospheres, iv) steam and gases, v) ammonia and hydrogen sulfide, vi) sulfur dioxide and oxides of nitrogen, vii) fuel gases, acids, alkalis and soils.

Therefore, the concept of corrosion in steels is then define as a natural electrochemical process that destroys the integrity of the metal structure in the presence of any environment containing moisture and oxygen. This process involves two electrode reactions that can occur in a spontaneously way at the interface between the metal and the aqueous environment according to the thermodynamic's Law; One, is the reaction of metal-base with chemical species from the environment (*i.e.* anodic-oxidation reaction, which discharge electrons from the metal substrate) and the second is the reduction reaction of an oxidizing agent (*i.e.* cathodic reaction, which restores the electron deficiency with reduction of protons from the metal surface). The exchange of electrons between anodic and cathodic reactions produces an electronic current flow across the metal interface, which is known as corrosion potential (E_{corr}). This means the value at which the two-coupled reactions are in equilibrium, some effects can be caused by imposing an electrical potential on the metal surface as much greater than the E_{corr} to favored the metal dissolution reaction as a soluble species that diffuses into the aqueous solution [4, 5]. This suggests that Fe contained in steel as a base component is oxidized and depends on the free energy like a driving force of E_{corr}. The transfer of the charge (ions/electrons) through the metal interface, react with the oxygen from the steel surface, with the subsequent growth of an unstable corrosion product in the form of a thick porous-oxide layer (also known as rust), which occupies more volume than the original material. However, hydrated iron oxides are not considered as a protective layer on steels in presence of negative ions, Cl^-, SO_4^{2-} or NO_3^{2-}. **Figure 1** shows a typical example of the degradation mechanism of concrete structures due to corrosion of the steel reinforcement embedded in it; i) initially, the pores of the concrete structure are the access pathway of negative ions that come from the environment, ii) then, corrosion reduces the cross-sectional area of the steel bar, iii) it produces oxides (hydrated ferric oxide-rust) with a larger volume that cause tensile stress in surrounding concrete areas, which results in cracking and subsequent structural failure of the concrete [6].

In other conditions, a thin oxide film can grow on metal-base to provide the protection against corrosion attack, that steels require in order to be useful when

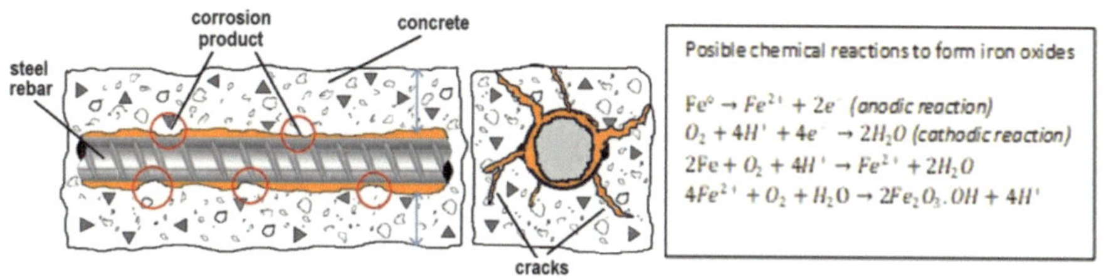

Figure 1.
Physical and chemical model that represents the concrete failure by corrosion on the steel reinforcement [6].

they are exposed to severe atmospheric conditions during their usage. This passive film is so thin that it is invisible to the naked eye; however, this film can be self-repaired immediately, when it is suddenly scratched or intentionally removed. An example of this outstanding property is the existence of several types of stainless steels that usually contains a significant proportion of chromium (12 to 25 wt.% Cr) with nickel and molybdenum to prevent the formation of ferrous hydroxide $(Fe^{2+} + 2OH^{-} \rightarrow Fe(OH)_2 \textit{ rusting product})$ by the presence of Cr in the Fe-base alloy, which reacts with oxygen from the environment to form a passive adherent oxide-layer (Cr_2O_3), thus, given a remarkably resistance to corrosion attack of the underlying metal, additionally, this oxide layer, can be regenerated by itself in the presence of oxygen [4, 7–9]. Based on the fundamental concepts, one of the advantages of using stainless steel is its high corrosion resistance, but in combination with other alloying elements can provide good mechanical strength, making the steel an appropriate material to be used in diverse applications that, in many cases, offers the only alternative for its high durability in aggressive environments; its use can be seen in domestic (cutlery, blades, household appliances and electronics), architectural (structures, handrails, concrete reinforcing bars, building components, cables for bridges and coastal works), transport (automotive exhaust system, ship containers, waste trucks and tankers for chemicals), chemical (pressure vessels, chemical containers, pipes, chemical plants, waste-water treatment), oil/gas (platform structures, machinery, storage tanks and pipelines), medical (surgical instruments, implants, equipment, dental inserts, wire and brackets in orthodontics), and other common uses (food containers, beverage bottles, springs, fasteners, bolts, nuts, washers and wires) [10].

For conventional steels produced by casting process, the most useful steel products are those that contain small amounts of alloying elements such as plain carbon steels (Mn, Si, S, P), alloyed steels (Cu, Ni, Cr or Al) and tool or machinery steels (W, Mo, Co, B and V). This alloying provides mechanical strength, ductility, machinability, and a substantial corrosion resistance. Although, these steels do not have the same ability of corrosion protection as the stainless steel does; the oxide film formed on the surface has only a few micrometers thick with microporous or growth defects, so it is possible to inferred that this oxide layer does not protect the metal from corrosion attack, this means a temporally low passivity is considered. However, in aggressive aqueous solutions the porous oxide layer can dissolve or break-down at least some areas of the film, therefore, leading to the Fe-base to a further localized attack. In industrial applications, the surface properties of the steel have a significant impact on their service life and performance. Among the several surface treatments to provide protection through a thick hard layer, diffusion techniques are using such as powder pack, gaseous atmosphere, plasma, ion beam and salt baths, that depends on the diffusion time and atmosphere concentration, these being a high effective treatment and less expensive. Additionally, carburizing, nitriding or boriding, are also well-known as thermochemical surface treatments [11–15].

Acid solutions are frequently used in many applications concerning industrial processes and are considered as the most corrosive media for steels. Acids like HCl, H_2SO_4, HNO_3, H_3PO_4, H_2CrO_3, and some alkalis such NH_3 are frequently used for surface cleaning, removal of rust deposits, pickling processes, chemical attack, metal surface treatments, and wastewater systems. Other relevant uses are metal-processing equipment, chemical processing, pipelines, food processing, chemical and petrochemical plants. Therefore, printed research works report several cases of using organic molecules compounds (imidazole, 2mercapto-benzimidazole, pyridine, thidiazole, pyrrolidine, triazole, among others) that have provided a significant corrosion inhibition property for steels during their exposure to acid media [16–24]. These molecules must contain in their structure functional electronegative groups, π electrons, heteroatoms or heteroatoms of nitrogen, sulfur and oxygen with aromatic and heterocyclic rings. These reports generally indicate that the molecules are dissolved in an ethanol-water solution and then added in small concentrations (ppm) to the acid media, in all the cases, a barrier layer of organic molecules is formed onto the metal surface by an adsorption mechanism, thus giving corrosion protection on steels under-service at aggressive conditions [16, 18, 20, 24].

According to Florian B. Mansfeld (1988) in his research *(Do not be afraid of electrochemical techniques —But use them with care)* [25] comments that corrosion is fortunately a problem that can be tracked by means of electronic devices (*i.e.* potentiostats) that applies an electrical signal (*V* or *I*) to measure and control the electrical charge transfer; in pursuance of evaluating the reaction kinetic and mechanism of corrosion process that takes place at the metal interface. Meanwhile, the constant improvement of measuring instruments and the availability of commercial software, makes possible an easy performance of the electrochemical tests for the evaluation of corrosion progress and its control in an experimental way. These achievements caught the attention of chemical, petrochemical, food processing and steel manufacture industries, as well as research laboratories and higher education faculties that have encouraged and certified the success of the use of electrochemical techniques to monitoring corrosion on steels. The application of electrochemical techniques, such as linear polarization, polarization resistance and potentiodynamic polarization, have often been used for several decades in evaluating successfully some basic phenomena as oxide passivity, effects of alloying elements, reaction kinetics and the use of inhibitors to control the corrosion behavior, among others. However, it is important to consider the limitations of the polarization techniques that use Direct Current (DC), to perturb the equilibrium of the interface between the metal and electrolyte solution, is the ohmic-droop that is often ignored, this occurs when the current flows through the resistance of the test solution and the resistance of the connecting cables to electrochemical cell electrodes (*i.e.* uncompensated resistance, *IR*) [25]. The effects of *IR* can cause severe distortions of polarization curves, leading in the erroneous estimation of corrosion rates and misinformation of the kinetic model that represents the potentiodynamic curves. Given this limitation, through the last decade, another electrochemical technique appears to be more suitable for corrosion studies, this is the Electrochemical Impedance Spectroscopy (*EIS*) that uses a small amplitude of alternate current (*AC*) in a certain frequency domain applied to the corrosion system under study. Usually, *EIS* data is collected through a potentiostat/galvanostat apparatus, and then it is fitted to an equivalent electrical circuit (EEC) model for its interpretation and analysis, fundamentally seeking a meaningful physical interpretation. In correspondence with several studies [26–33] *EIS* is considered a successful new electrochemical technique with a great evolution in recent years that has become an essential analytical tool in the research of materials science. For its detailed information, versatility and sensitivity that makes possible to be used widely in; corrosion studies and corrosion control, monitoring of properties of electronic and ionic conducting polymers or ceramics, colloids and coatings, measurements in semiconductors and solid electrolytes, studies of electrochemical kinetics at electrode-media interfaces, determination of conducting or diffusion mechanism, reactions and process [34].

The practical estimation of *EIS* technique could be difficult to understand by non-specialist because of the lack of comprehensive and explanation about the theory's basic aspects in conjunction with the experimental measurements. Nevertheless, it is possible to attain a logical interpretation and analysis of acquired impedance data for a certain practical system, such as those studied in this chapter that will be shown later. In this sense, to avoid uncertainties and misinterpretation of impedance data, analytical co-relationship of physical, chemical and manufacture parameters must be established with an equivalent electrical circuit (EEC) model, thus given a common sense to the impedance response. Therefore, this review considers a wide variety of practical electrochemical impedance cases for the study of corrosion mechanism on steels based on the basic aspects of *EIS* theory and its experimental interpretation. This chapter serves as a support for postgraduate students to have a criterion in deciding through their own experiences when using the electrochemical impedance technique. The practical cases discussed here are part of the research experienced by Dr. Héctor Herrera Hernández known in the scientific community as *DR.3H*. Recently, *DR.3H* and his students & research group are dedicated to their experience in electrochemical impedance knowledge in medical applications as bone replacement or PVDF-based membranes as an appropriate scaffold for skin cell growth [35].

Cases of *EIS* study applied to steels;

- Steels measured in their received condition.
- Kinetic oxidation reaction at different aqueous solutions.
- Steel corroded at non-stationary condition.
- Corrosion monitor in concrete reinforced materials.
- Inhibition using organic molecules.
- Inhibition in natural liquids extracted from plants.
- Hard-coatings as protection.
- Corrosion monitor in steels used as food containers or beverages.

1.1 Fundaments of electrochemical impedance spectroscopy (EIS)

Since the middle of the 18th century, the Impedance Spectroscopy (IS) technique has been established as a popular theoretical approach to study the electrical properties of conducting materials and their interfaces. However, in the last quarter-century, IS becomes a practical tool that is successfully applied in electrochemistry as an analytical method widely used in many disciplines such as materials science, corrosion technology, semiconductors, conducting polymers, ceramics, coatings, energy storage, and solid-state. Electrochemical Impedance Spectroscopy (EIS) is considered as a new technique with astounding advantages [36–38].

The concept of impedance in electronic devices is generally treated as a purely complex phenomenological amount and is considered as one of the most important physical characteristics that concerns the resistance that the medium opposes to the propagation of sound (acoustic impedance, *Z*) through it and therefore it is equivalent to the electrical impedance. In this sense, acoustic impedance is the ratio of the sound pressure of the wave (P) to its volume speed (U) in a material medium [39, 40]. This concept becomes a similar analogous meaning to the electric approach, because an electrical impulse (*V* or *I*) is applied to the conducting electrodes and a characteristic electrical response is resulted, known as impedance, *Z*. Therefore, impedance is then defined as the measure of the ability of a certain circuit to resist the flow of electrical current. The electrochemistry impedance is the relationship between a potential energy difference and the flow of electrons generated by a wave signal applied in an aqueous media. *EIS* technique is characterized by using an alternating current (*AC*) signal as driving force, which is applied to a conductive electrode, thus obtaining a characteristic response from the system interface. One of the attractive aspects that makes *EIS* as a suitable tool for investigating the electrochemical properties of materials during their exposure to aqueous solutions, is the simulation of the system behavior by means of an idealized circuit model. This consists of an arrangement of passive electrical components (*i.e.* resistors *R*, capacitors *C* and inductances *L*), which are the physical representation of the electrochemical processes occurring at the system interface under study.

Another quality of *EIS* is its high measurement sensitivity, which makes the technique an attractive advantage for detailed information that can be obtained from the system in study. For example, *EIS* was used to evaluate the properties of thin oxide films formed on metals, monitoring superficial degradation of polymer layers or paint coatings due to swelling process (coatings damaged by water uptake). Surface changes due to ion adsorption at the interface can also be detected, knowing the kinetics reaction on metals under corrosion process; all this, due to the advantages of this technique to perform measurements using a very small amplitude signal at variable frequency range. As result of the advantages mention above, EIS has attracted the interest of many scientist and engineers from different areas of application, for example: corrosion technology, electrochemistry, metallurgy, hydrodynamic, chemistry, biology, physics, mechanical, and medicine. According to organic chemistry a molecule is a group of non-electrically charged particles that have two or more atoms chemically bonded. They are components of the matter lying on earth (minerals, atmosphere, gaseous substances, organic and inorganic compounds, liquids, among others) [41]. Molecules can be measured with a small *AC* amplitude of voltage as a function of the frequency without altering their properties. Some systems leading to the formation of interfaces with the materials for example; a solid–solution interface allows the ion charge transfer, conduction and electron flow that is governed by the free energy of the chemical reactions occurring at the interface region (named double layer), as is shown in the model of **Figure 2**. The electrical properties of the double layer can be measured by an electrical equivalent circuit, considering that the double layer behave as pure capacitor C_{dl} (ions charge) and the flow of ions through the metal surface is view as a resistance R_{ct} of current, in according to Ohm's law. In general, *EIS* allows separating the contribution response of different components in terms of the resistance of electron charge transfer, double layer capacitance, solution resistance, inductance, and other parameters, where several electrochemical processes are proceeding at a different reaction rate.

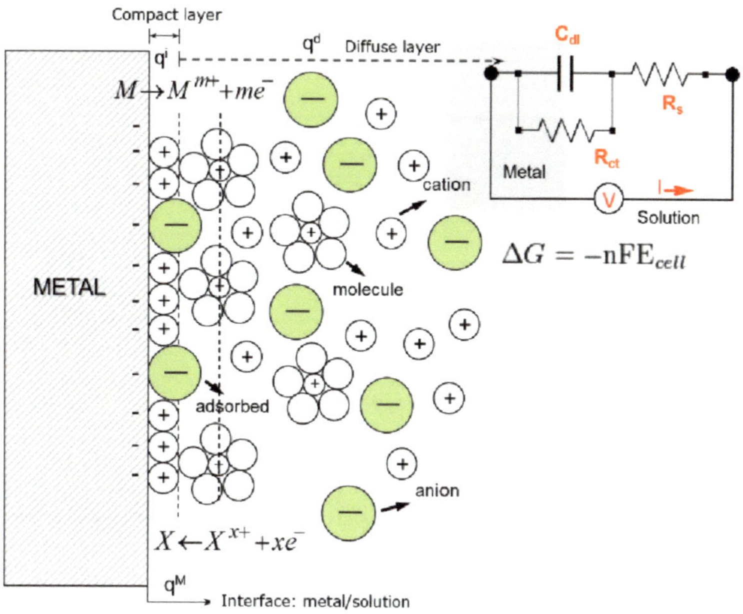

Figure 2.
Schematic simulation of the electrochemical interface metal/electrolyte (electrical double layer) and its representative equivalent electrical circuit EEC *model. R_s is the solution resistance, R_{ct} is the charge transfer resistance, C_{dl} is the capacity of the double layer.*

1.2 Basics aspects of EIS data representation

Electrochemical impedance spectroscopy (EIS) is the analytical method widely used to study the electrochemical systems by applying a small *AC* voltage signal as a function of frequency of the amplitude signal. In potentiostatic mode as that of direct current (*DC*) techniques, like Linear Polarization Resistance (*LPR*) or Polarization Potentiodynamic, the basic measurement parameter is the polarization resistance R_p that is equal to the impedance (*Z*) in alternate current (*AC*) mode. This can be represented according to the Ohm's Law equation as denote bellow [8]:

$$R = \frac{V}{I}(DC), Z = \frac{E}{I}(AC) \tag{1}$$

where *R* is the resistor (Ω), *V* is the voltage (volts) and *I* is the current (amps) for direct current and *E* is the potential (volts) and *Z* is the impedance (Ω) for alternating current. To understand how the theory supports the EIS technique, it is necessary to consider two periodic waves; one is the current signal (*I*) and the other is related to potential signal (*E*). These waves behave as that shown in **Figure 3**, in which both signals oscillate at the same frequency and intensity, because one wave causes the other. However, there is an important effect that is the constant time shift between the two waves at certain angle, this is called the phase-angle shift ($\boldsymbol{\phi}$) and can vary from 0 to 90. Its unit is degrees (°), because usually waves are considered vectors in a polar coordinate system or in a sine function. **Figure 3** shows the relation between waves *E*, *I* and the phase-angle shift. The applied sinusoidal perturbation can be a potential signal (*E*), thus given the measurement response in current (*I*) at a certain frequency domain. The excitation signal as a function of time *t* is represented as follows;

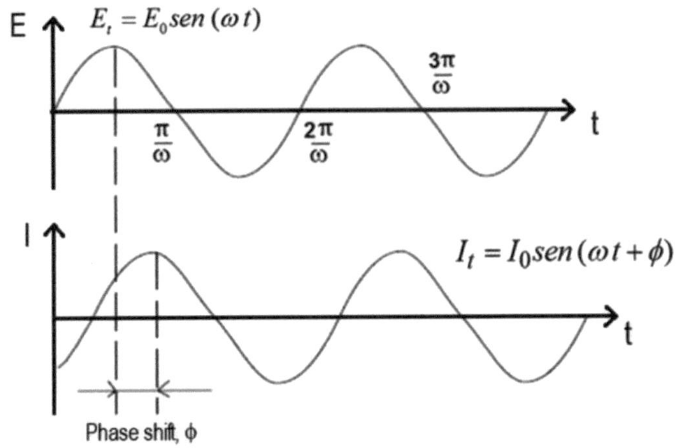

Figure 3.
Sinusoidal waveform response in linear system showing phase-shift angle that is used to describe the electrochemical reactions at the interfaces [42].

$$E_t = E_o \sin(\omega t) \tag{2}$$

where E_t is the potential at time t, E_0 is the amplitude of the signal and ω is the angular frequency (expressed in terms of radians/second). So, the relationship between angular frequency and frequency (f in hertz units) is given by;

$$\omega = 2\pi f \tag{3}$$

in order to preserve the linear behavior in electrochemical systems, a small amplitude of AC voltage of about 5 to 10 mV is usually applied. I_t *is* the single response of instantaneous current at the maximum amplitude, Φ is the shifted-phase angle and has a different amplitude, I_0 as described in Eq. 4;

$$I_t = I_o \sin(\omega t + \phi) \tag{4}$$

taking into account the electrical parameters of E and I as a function of angular frequency in the time domain, as well as the shifted-phase angle is possible to rearranged the Eq. 2 and 4 into Ohm's Law as DC current, this relationship enables to calculate the impedance of the system under study as follows;

$$Z = \frac{E(t)}{I(t)} = \frac{E_0 \sin(\omega t)}{I_0 \sin(\omega t + \phi)} = Z_0 \frac{\sin(\omega t)}{\sin(\omega t + \phi)} \tag{5}$$

then, impedance (Z_0) is defined as the ratio of applied voltage (E) divided by current (I) and represents an opposition to the flow of electrons or current in an AC circuit due to the presence of resistors, capacitors and inductors. Among of variety of passive electrical components, only resistors and capacitors or inductors contribute manly to impedance; one is related to the real component (Z') and the other to the imaginary component (Z''). Due to this assumption, Z_0 can be expressed in its complex notation by incorporating the complex number $j = \sqrt{-1}$, where the **Figure 4** shows the complex representation of the impedance as vector concept, $Z(\omega) = Z' + jZ''$ and its phase-angle, $tan\ (\phi) = \frac{Z''}{Z'}$. Using Euler relationship, the expression of the impedance translates in a complex function like;

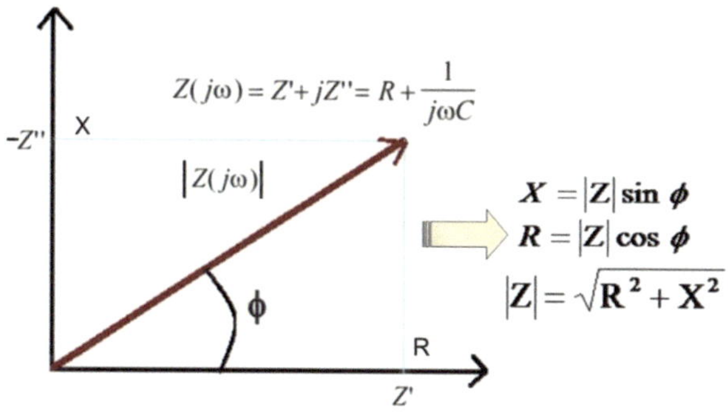

Figure 4.
Vector representation of impedance as complex number; X *capacitive-reactance,* R *resistance, Z' real component and Z" imaginary part of impedance.*

$$E_t = E_0 \exp(j\omega t), I_t = I_0 \exp(j\omega t - \phi) \tag{6}$$

considering the relationship between the potential and current amplitude, it results the total of the impedance as a complex number, as it follows;

$$Z(\omega) = \frac{E}{I} = Z_0 \exp(j\phi) = Z_0(cos\phi + jsin\phi) \tag{7}$$

however, the use of the current as a perturbation signal is also designed for certain electrochemical applications. Once the experimental data are collected, a series of potential-time and current-time are obtained, which correspond to the impedance at each frequency studied. The representation of the *EIS* data is by means of Impedance Spectra known as Nyquist Plots ($-Z_{imag}$ vs. Z_{real}) that represent the real impedance plotted against its imaginary part and also is often used the Bode plots (*log*|Z| vs. *log f*req., ϕ vs. *log f*req.) that is the graphical representation of the modulus of the impedance and its phase-angle, as a function of the frequency domain) [43, 44]. However, Mansfeld [25] is his reports suggest that Bode plots is more appropriate to represent the impedance data because most of the measuring points are completely displayed at the entire frequency domain of the spectra. Thus, allowing a quick diagnosis of the behavior due to the sensitive of the phase-angle to small changes as a function of frequency variation, expecting time constants. While, Nyquist's diagrams are not recommendable since most of the data is grouped together at both ends of the spectra.

Experimentally speaking, when an *EIS* analysis is chosen to study the corrosion behavior of a piece of metal (*WE-working electrode*) that is immersed in an aqueous solution for a certain period of exposure time, which its equilibrium is perturbed by a low amplitude sinusoidal signal as function of frequency in the presence of a polarizable counter electrode (*CE*) and a reference electrode (*RE*), it is necessary to consider some electrical parameters (*i.e.* dielectric constant, permittivity, conductivity, resistivity and capacity charge) that will allow to interpret and deduce the corrosion behavior and its reactions mechanism by modeling the *EIS* data to an electrical *RC* circuit. These *RC* circuits are assembled with capacitors (*C*) and resistors (*R*) in parallel or series. C_{dl} is used to represent the electrical charge transfer at the metal/electrolyte interface known as the capacitance of a double layer (in farads), and that is present in all corroding aqueous systems. R_{ct} is the resistance of the electron charge transfer, which is the value of the impedance in its real component and R_s is the solution resistance. The combination of these three passive elements provides a simple equivalent electrical circuit (*EEC*) for a uniform corroding metal. The experimental contribution of each parameter mentioned above is like that shown in **Figure 5**.

1.3 Analysis and interpretation of EIS spectra

As mention above, *EIS* data is usually represented by Bode plots in which the $|Z|$ module and phase angle ϕ are fuctions of frequency domain, sustained by its complex plane form that relates to Z_{real} with the imaginary part Z_{im}, and are usually interpreted by a mathematical correlation to a certain physico-electrical model known as equivalent electrical circuit (*EEC*), which is designed by an arrangement of ideal components (resistors R, capacitors C and inductors L) connected in series or parallel in order to reproduced the experimental *EIS* spectra. The impedance values of these elements are associated to the electrochemical processes of real systems such as electron charge transfer, diffusion processes, determination of the capacitance of the electrochemical double layer, mechanism of ions adsorption, mass transfer kinetic, characterization of coatings integrity, electrical resistance of the electrolyte, corrosion detection, conductivity, electrochemical reactions, among others, **Table 1** shows different *EEC* models that were designed to simulate & interpreted in particular some of the most common electrochemical processes, which will help to understand and describe the *EIS* spectra obtained during an experimental procedure. For example, If an alternating voltage $E(t) = E_o * Sin(\omega t)$ of about 10 mV at 1 Hz is applied to the *RC* circuits that are shown in **Table 1** as the perturbation energy of the models to carrier electrons through their passive components, this results in a signal that has a sinusoidal behavior and varies as a function on time (*i.e.* current intensity $I_o = \frac{E_o}{R}$), this waveform moves in the same direction and frequency as the supplied potential. However, to simplify the use of sinusoidal signals and their effect on different electrical components such as *R* (resistor), *C* (capacitor) and *L* (inductor), the typical sinusoidal response of the *RC* components is like that are shown in **Table 2**, and also it shows their relation to the shifted phase-angle value, the impedance as a function of time-frequency and their relationship to the electrochemical processes.

A single *RC* circuit is first considered to have only one ohmic resistance of 3.3 kΩ connected to a power source, in this case, the current intensity flows constantly through the resistor without any phase difference with respect to the potential that originates the waveform signal, $\phi = 0°$. Then for this condition in that the phase-angle is equal to zero, the value of impedance module for a pure resistor (*R*) is relatively its reactive or real part ($Z_R(t) = R$), being its imaginary part or the

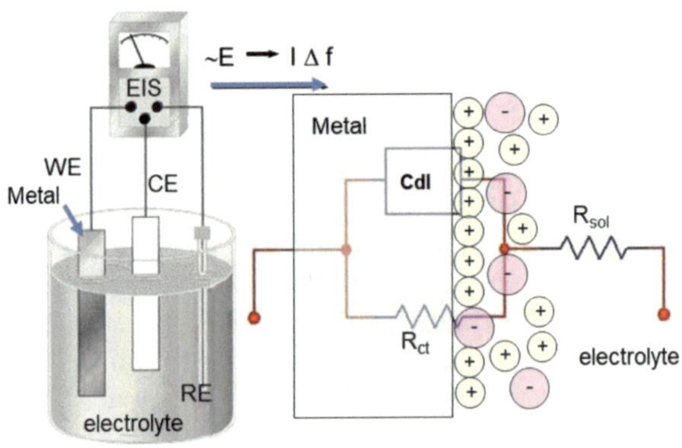

Figure 5.
*Representation of a corrosion cell and its equivalent electrical circuit (*EEC*),* WE is the *working electrode,* CE *is the counter electrode and* RE *is the reference electrode.*

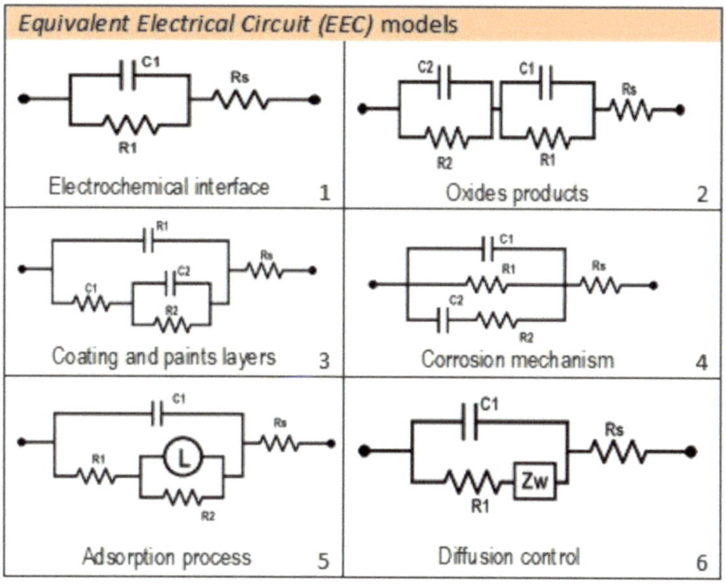

Table 1.
EEC models used to describe the electrochemical processes most studied by EIS. *1. Electrochemical interface (electron transfer), 2 and 3 oxide products and coatings, 4. corrosion mechanism, 5. adsorption and 6. ion diffusion processes.*

Component	Description	Impedance, Z	Waveform signal
Resistor (R)	Is use to regulate the current flow trhough a particular path and obeys the Ohm's law (Ω). $E_{(t)} = V_o Sin\ (\omega t + \varphi_v)$	$Z_R = R$ $\varphi_i = \varphi_v$	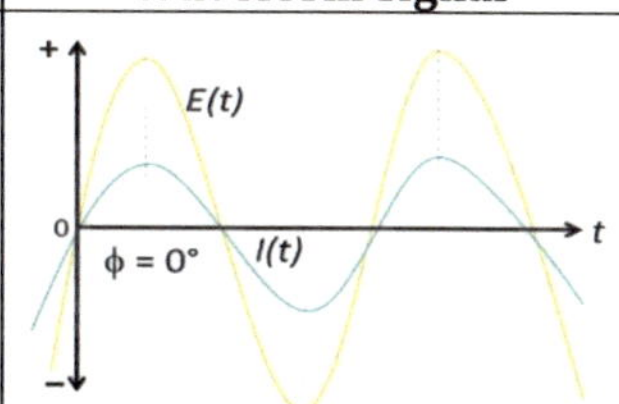
Capacitor (C)	Component with capacity to store electrical charge q(t) in Faradios (F). $E_{(t)} = V_L = L\frac{d}{dt}\ [i(t)]$	$Z_C = -\frac{1}{j\omega C}$ $\varphi_i = \varphi_v + 90°$	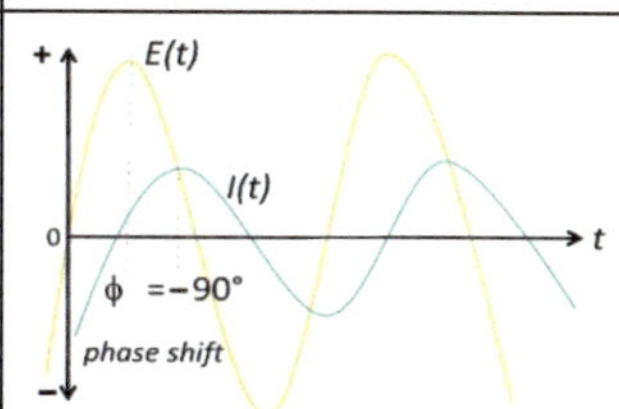
Inductor (L)	Coil of wire that induces a magnetic field as a result of current passing in henries (H). $E_{(t)} = V_C = \frac{1}{s}\int i(t)\,dt$	$Z_L = j\omega L$ $\varphi_i = \varphi_v - 90°$	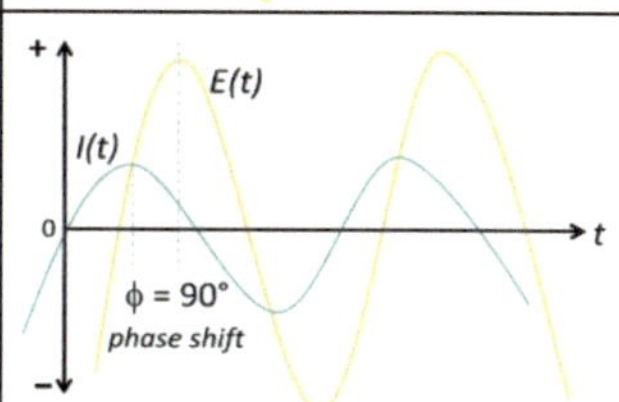

Table 2.
Impedance and phase shift angle response for the passive RC *electrical components.*

reactance X equal to zero, so it is suggested that this behaves as a resistive component. The impedance diagram for a resistive component shows a single straight line on its real axis that tends to infinity as dependent on time and frequency domain. In the case of a pure capacitor (C) the sinusoidal response of voltage is retroceded at least by $-90°$ allowing the imaginary component to be the variable dependent on time and frequency domain, so its value of $Z_C(t) = -\frac{1}{j\omega C}$. In the opposite case, it happens for an inductor (L) in which the current waveform signal is advanced near to 90°, which gives the expression $Z_C(t) = j\omega L$.

On the other hand, when two passive components are combined in a RC circuit, for example, one resistor of about 276 Ω and a capacitor of 1 μF which are connected together in series, a small electrical AC signal of 10 mV is supplied to flow electrons through the closed circuit as dependence of frequency domain from 1 MHz to 1 mHz, the impedance is given by $Z_T(t) = -\frac{1}{j\omega C} + R_o$, and depending on the resistance and capacitor values can be considered as capacitive or resistive behavior. The Bode and Nyquist plots of **Figure 6a** show the experimental simulation of impedance response for these RC components connected both in series, which behave like a capacitive. Circuit model #1 shows the simplest arrangement of series and parallel, in which a resistor of 3.3 kΩ is connected in parallel to a capacitor of 1 μF and then in series with other resistance of 276 Ω, its impedance response could be described as a function of frequency according to the following equation $\left(Z_T(t) = \frac{R_1}{1+(j\omega C_1 R_1)} + R_o\right)$, see **Figure 6b** the corresponding impedance spectra. So, circuit #1 is known as *Randles* circuit and is the typical electrical model used to described as analogy form the physical phenomenon of metal under corrosion attack [45, 46] by electron charge-transfer at the interface metal/electrolyte, and also to simulate uniform corrosion on homogeneous surface, which it has been the most used along the decades on researches. The information data of the impedance spectra is clearly visible in **Figure 6b**, because it is possible to obtain the value of the frequencies corresponding to the solution resistance R_s, charge transfer resistance R_{ct}

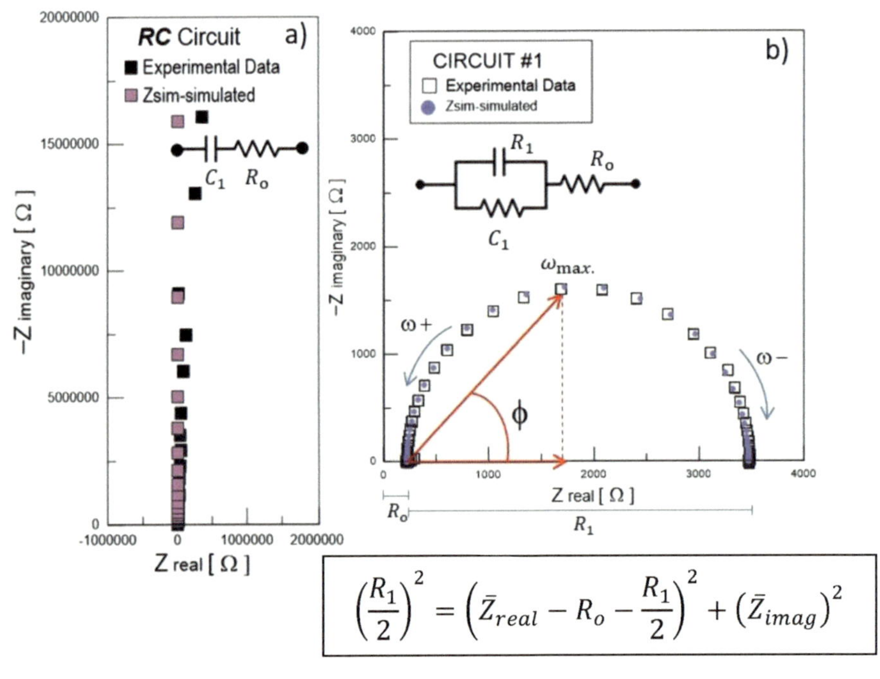

$$\left(\frac{R_1}{2}\right)^2 = \left(\bar{Z}_{real} - R_o - \frac{R_1}{2}\right)^2 + \left(\bar{Z}_{imag}\right)^2$$

Figure 6.
Impedance data simulation of a simplest EEC model; a) RC *elements in series (high coating impedance) and b) circuit #1* RC *in parallel follow by* R *in series (simple corrosion undergo by electron charge transfer).* $R_o = 276\ \Omega$, $R_1 = 3.3\ k\Omega$, $C_1 = 1\ \mu F$.

(or polarization resistance), and its capacitance of the double layer C_{dl}.
From the Nyquist Plot of **Figure 6b** C_{dl} is clearly observed as a well-defined semi-circle (*a single time constant*) in the entire frequency domain as results from the electrical *circuit #1*, which is the diameter of this semicircle is equal to R_{ct} and R_s is obtained by reading the real axis, Z', value at the high frequency intercept. However, considering the maximum angular frequency (ω_{max}) as the frequency at which the imaginary component of the impedance Z'' has its largest value and R_{ct}, the value of C_{dl} is given by the following expression;

$$C_{dl} = \frac{1}{\omega_{max} R_{ct}} \tag{8}$$

Two-time constants could be expected in **Figure 7a** (circuit #2 is the combination of parallel *RC* in series) $\left(Z_T(t) = \frac{R_2}{1+(j\omega C_2 R_2)} + \frac{R_1}{1+(j\omega C_1 R_1)} + R_o\right)$, or **Figure 7b** (circuit #3 is a parallel arrangement in parallel connection $\left(Z_T(t) = \left\{\left[\frac{R_2}{1+j\omega C_2} + R_1\right] \| \frac{1}{j\omega C_1}\right\} + R_o\right)$ as results of applying voltage through the circuits or **Figure 7c** (circuit #4) $\left(Z_T(t) = \left[\left(R_2 + \frac{1}{j\omega C_2}\right) \| R_1 \| \frac{1}{j\omega C_1}\right] + R_o\right)$. These *EEC* models are used to describe the impedance spectra for oxide products forming by corrosion reactions on the metal surface, or anodizing coatings, or for paint-coated metals after exposed to corrosive electrolytes. Where C_1 is the capacitance of oxide film connected in parallel to R_1 that is the oxide resistance, both connected in series to *RC* parallel that contributes the electrical response of inner barrier layer or the double layer interface (oxide/metal matrix). The capacitance value of coating is measured in Farads [F], which depends on its dielectric constant ε and its thickness d as given by;

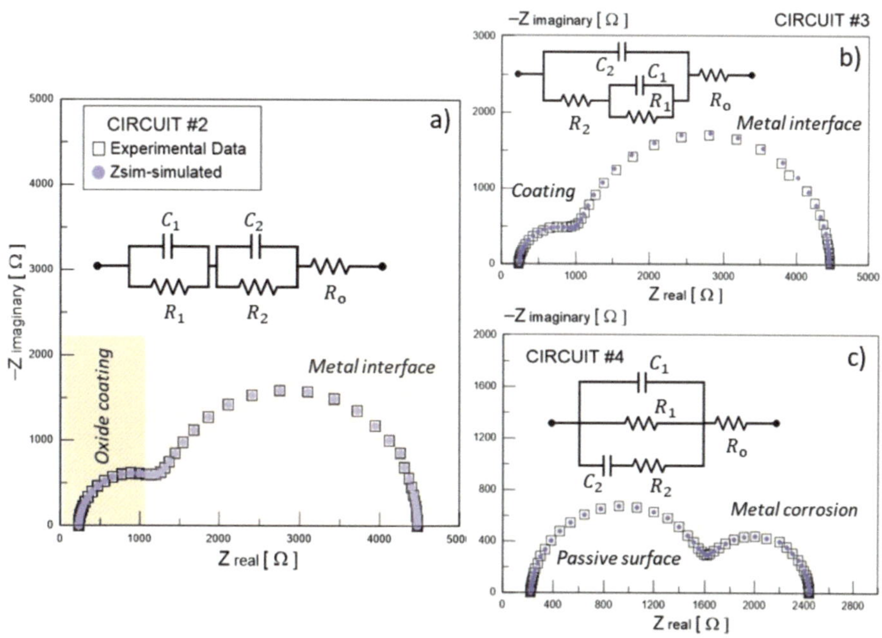

Figure 7.
Impedance data simulation of EEC *models; a) circuit #2 parallel* RC *elements in series (oxides products), b) circuit #3 parallel* RC *elements in parallel follow by* R *in series* (paint coating) *and c) circuit #4 parallel* RC *elements in series (corrosion mechanism).* R_o = 276 Ω, R_1 = 3.3 kΩ, C_1 = 1 μF, R_2 = 1 kΩ, C_1 = 10 μF.

$$C_c = \frac{\epsilon_o \epsilon A}{d} \tag{9}$$

Where ϵ_o is the electrical permittivity constant of free space (8.85×10^{-12} F/m) and A is the exposure area of the coated electrode. So, it is expected that the capacitance of coating increases with the exposure time due to water up-take by the coating through ionically conducting paths called pores. Changes in pores resistance and capacitance can be used to estimate the corroding metal.

Other types of impedance spectra commonly observed in the printed research works, are similar to that reproduced with the simulation using circuit #5 or #6. Circuit #5 is the similar arrangement of circuit #3, in which the ideal capacitor C is replaced by a magnetic coil (inductor) $L\left(Z_T(t) = \left[\left(\frac{j\omega R_2 L}{R_2 + j\omega L} + R_1\right) \| \frac{1}{j\omega C_1}\right] + R_o\right)$. In this sense the impedance diagram in **Figure 8** shows a semicircle very well defined by its diameter throughout the frequency range (charge transfer process) but is accompanied by a second inductance response below the semicircle at low frequencies, that means adsorption ion mechanism. This impedance response is commonly observed in electrochemical systems where chemical species, ion or any molecule is physically adsorbed at the interface of the electrochemical double layer with a given electrical charge motion. While circuit #7 is derived from the simplest circuit #1 in which its resistor is replaced by another electrical element $Z_w\left(Z_T(t) = \frac{R_1 + W}{1 + [j\omega C_1 (R_1 + W)]^a} + R_o\right)$ called Warburg impedance and related to the diffusion control of the species this can happen when the surface concentration of an electrochemically active species changes during the AC cycle. Thus, it must consider the impedance of a cathodic reaction, such as the reduction of oxygen that is common in corrosion systems. The general shape of a Warburg impedance is shown in **Figure 8b**. Two regions are clearly seen; a semicircle response is due to the charge transfer reaction and straight line with a 45° angle to the abscissa means to the diffusion of reactants [6, 43–45]. This is typical for analytical electrochemistry in diffusion controlled (W) in corrosion measurements, which is expressed by the Eq. (10), where σ is the Warburg coefficient and can be calculated from the slope of the straight line in the complex plane of **Figure 8b**.

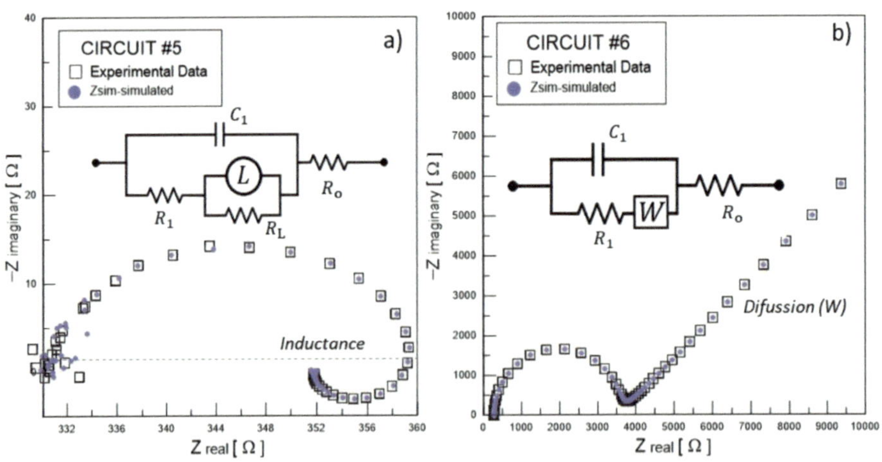

Figure 8.
Impedance data simulation of EEC *models; a) circuit #5 parallel* RC *elements in parallel with an inductance* L *(adsorption mechanism) and b) circuit #6 parallel* RC *elements in series with Warburg impedance* W *(diffusion control).* $R_o = 276\ \Omega$, $R_1 = 3.3\ k\Omega$, $C_1 = 1\ \mu F$, $W_o = 0.001\ S\text{-}sec^{0.5}$.

$$W = \sigma\omega^{-1/2(1-J)} \tag{10}$$

In real cases the shape of Nyquist plot does not always show a perfect semicircle as it is observed for pure capacitor, it is necessary to replace capacitor (*C*) by a Constant Phase Element (*CPE*) in order to compensate the depression of the semicircle of frequency dispersion resulting of an experimental system due to the surface inhomogeneity, surface roughness, electrode porosity, surface disorder, geometric irregularities, and others. The *CPE* is a mathematical expression that is useful to represent several electric elements $\left(Z_{CPE} = \frac{Q_o}{(j\omega)^{1-\alpha}}\right)$, [47]. To obtain the capacitance value (C_{dl}) from the *CPE*, it is necessary to obtain the maximum frequency of the Nyquist semicircle ($\omega_{\theta max}$) as well as the *n* exponent, this exponent can have values between 0.7 to 0.9, which can be used to describe the experimental data and a physical meaning is not yet clear, Q_o is a constant element with dimensions S-sec^n. Eq. 11 shows the calculation of C_{dl}:

$$C_{dl} = Q_O * \omega_{\theta max.}^{(n-1)} \tag{11}$$

In **Figure 9** is shown the configuration of *EEC* for a Nyquist Plot obtained experimentally from a corrosion system, the use of *CPE* was useful to adjust the experimental data to a mathematical fit in order to obtain the corrosion behavior of the metal (carbon steel APIX-52-5 L) in acidic media) HCl1M [20, 47].

The validation of the parameters obtained through an analogous *EEC* model can be evaluated through the Kramers-Kroning Transformations (*KKT*), this is done in order to evaluate and understand the mechanisms that occur in the system interface. *KKT* are mathematical relationships between the real and the imaginary parts of the impedance that must be obeyed by valid impedance data. Therefore, meaning

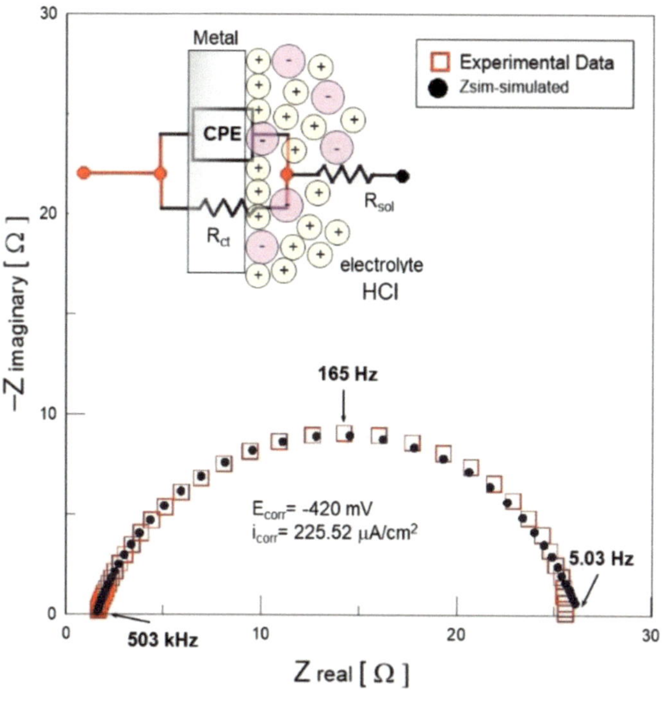

Figure 9.
EEC *electrical circuit #1 with a* CPE *to fit de impedance data corresponding to corrosion of pipeline steel immersed in HCl 1 M.* R_s = *1.58 Ω-cm²*, C_{dl} = *53.04 µF/cm²*, R_{ct} = *24.71 Ω-cm²*.

that when imaginary impedance Z′′ is known for all frequencies, it is possible to calculate the real impedance Z′ at all frequencies [48, 49]. The general conditions on which *KKT* are based are show bellow:

1. Causality. The response of the system is due only to the perturbation applied and does not contain significant components from spurious sources.

2. Linearity. The perturbation and response of the system are linearly related *i.e.* the impedance is independent of the amplitude of the perturbation signal.

3. Stability. The system must be stable in the sense that it returns to its original state after the perturbation is removed.

4. The impedance must be finite-valued at $\omega \rightarrow 0$ and at$\omega \rightarrow \infty$ and must be continuous and finite-valued function at all intermediate frequencies.

It has been shown that when a corroding system obeys the just mentioned four criteria the impedance data will converse correctly. However, the inverse is not always true. It is still possible to have a correct *KKT* when impedance data are nonlinear. In the case of our impedance measurements we are mainly concerned about the stability of the system and for this case the *KKT* is an excellent tool for data validation.

2. EIS applied to metal corrosion

2.1 Analysis of effects of a voltage stimulus applied to corroded metals

One of the principal applications of *EIS* is in the study of electrolyte/electrode interfaces which is widely used in the evaluation of corrosion mechanism in metals at different environments conditions, but it has also been very useful in the performance of coatings [50–55] and in the failure detection of materials by stress corrosion cracking, similarly according to recent publications *EIS* also appears to be applied in ceramics materials [56–58]. In this sense, most of literature indicates that when applying a periodic signal of potential with amplitude from 5 to 10 mV in a given frequency domain, it is possible to detect the transitory current to obtain a change in the phase angle between *I-V* and the $|Z|$ data, which progress over time in order to predict metal corrosion phenomena or a possible electrochemical reactions at the metal interface. It should be noted that using a known electrical circuit it is possible to characterize the impedance spectra for each system under study as it shown before. The device that allows applying a programmed potential and detected the current is a potentiostat. Therefore, in this study a galvanostat-potentiostat PARSTAT-4000 was used to evaluate the effect of the voltage applied to the two-electrode interface. In which a periodic constant signal at 1 kHz of frequency was applied over a voltage range of 1 to 1000 mV as a function of frequency domain (1 MHz to 1 mHz). For this study it was considered the following systems; i) An ideal system like circuit #1, which is designed by *RC* components, a pure capacitor of 1 μF is connected in parallel to a resistor of 3 kΩ and then connected together in series with a resistor of 200 Ω and ii) a 3 cm^2 of stainless steel plate were used as working electrode (*WE*) after being exposed to an aqueous solution of HCl 1 M, then the *WE* was perturbed by a sinusoidal potential at different amplitude from 1 to 1000 mV, the corresponding impedance data for each of the cases that are displayed in **Figure 10**.

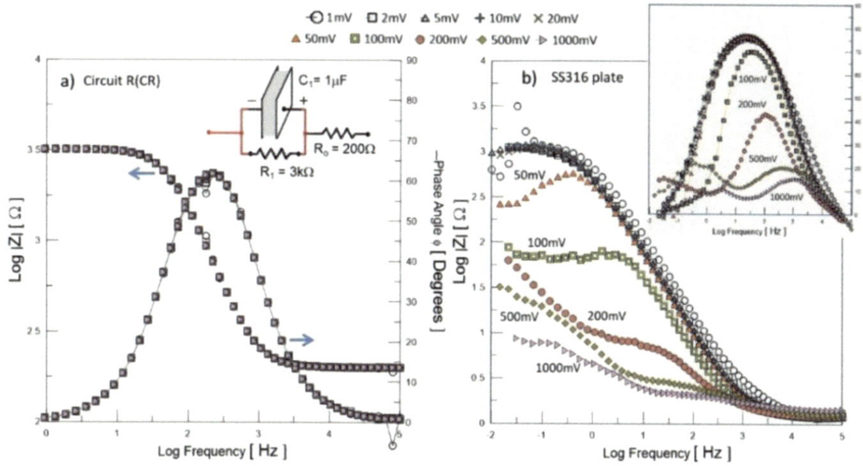

Figure 10.
Typical impedance spectra showing the effects of the amplitude signal in; a) EEC *model #1 (*$R_o = 276\ \Omega$*,* $R_1 = 3.3\ k\Omega$*,* $C_1 = 1\ \mu F$*) and b) a stainless steel immersed in HCl 1 M.*

The results show that when an alternate electrical pulse *V(t) of* 1 kHz fluctuates from 1 to 1000 mV through an ideal circuit like *EEC* model #1 as that shown in **Figure 10a**, a uniform current *I(t)* flows as a function of frequency domain, this signal produces a well-defined time constant in the entire frequency range. During the pulse at a time *t* the capacitor stores electrical energy causing an increase in potential difference $\left(Z_T(t) = \frac{q}{C}\right)$ and that allows the current to be phase shifted with respect to the voltage of about 60°, meanwhile the resistor R_1 connected in parallel does not allow the passage of the current, instead of it decreases gradually to zero according to the Ohm's Law, that is why the capacitor stops charging load. Finally, when the period of the capacitor's transient load ends, the potential difference in the circuit must be zero when the stored load has been exhausted, *i.e.* the circuit has been returned to its equilibrium state. Due to the characteristics of the capacitor, which is composed by a parallel polished metal plates separated with a dielectric at a distance of *d*, and due to the transient events of charging rate and discharging rate during the continuous passage of the potential at different intensities of the signal amplitude does not cause changes in the interface of the plates, so the impedance data in bode representation are overlaid showing the same behavior for all data. That is, the load capacity or capacitance of 1 μF remains constant as the amplitude of the sinusoidal signal increases from 1 to 1000 mV as is shown in **Figure 11**.

The same behavior is observed for stainless steel SS316 plate immersed in HCl 1 M (**Figure 10b**), the metal interface exposed to the acid solution allows the electron transfer rate at the equilibrium potential (E_{corr}) after applying lower amplitudes of the stimulus signal (between 1 to 20 mV), the impedance diagrams for this conditions do not show changes caused by the current flows into the system. In this sense the metal interface working similar as the ideal capacitor allowing ions loading charging such as Cl^- and OH^- with capacitances ranging between 40 to 80 $\mu F/cm^2$, which is indicated by a well-defined one time constant due to the presence of a protective oxide layer (passive condition) and can be easily represented by the *EEC* model #1. Notable effects can be caused by applying high current, as is clearly seen in the distortion of the shape of EIS diagrams during increasing the amplitude of the stimulus signal from 50 to 1000 mV, the impedance value $|Z|$ gradually down several orders of magnitude and severe changes in phase angle less than 20° are observed, this mean that two time constants are obvious seen and are related to the corroded interface, *i.e.* dissolution of the chrome protective

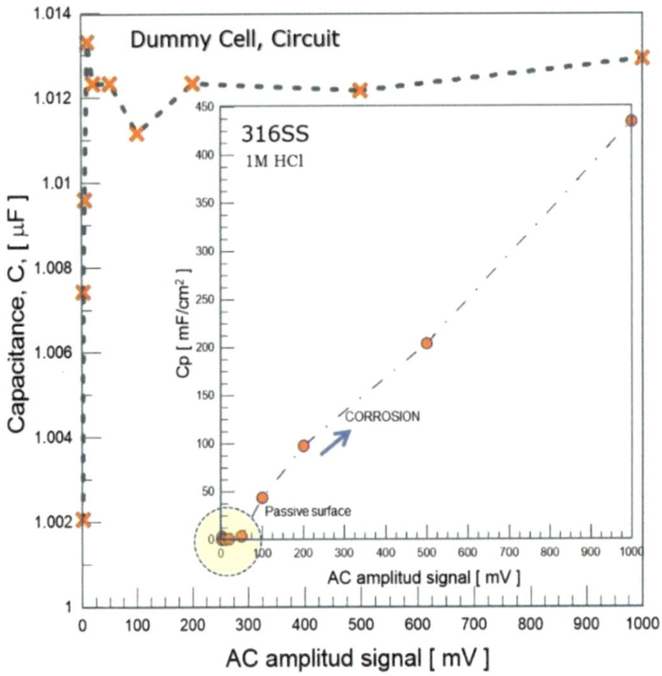

Figure 11.
AC amplitude signal dependence on capacitance value for an ideal EEC *circuit model #1 (C_1 = 1 μF) and the stainless steel SS316 plate during its immersion in HCl 1 M.*

film and manifestation of the pitting corrosion process that occurs after 200 mV, for this case an increase in the interface charge of electrons is expected with capacitances over 434.40 mF/cm^2, like that as shown in **Figure 11**. It can conclude that it is possible to carry out experimental tests with amplitude signals ranging from 1 to 20 mV at the steady-state of corrosion potential without surface damage by the current applied, which is in according to the literature that reports an amplitude signal of 5 to 10 mv.

2.2 Kinetic oxidation reaction of steels tested in their received condition at different aqueous solutions

Figure 12 shows the typical impedance behavior of a steel with specification of AISI 8620 (0.20 wt.%C, 0.90 wt.%Mn, 0.35 wt.%Si, 0.60 wt.%Cr, 0.70 wt.%Ni, 0.25 wt.%Mo) in its received condition after exposed to different aqueous solutions such as distilled water, NaCl 0.5 M, HCl 1 M, H_2SO_4 1 M. The supplied voltage signal has an amplitude of 10 mV that fluctuating around the corrosion potential (−654 mV) in the frequency range of 1 MHz to 1 mHz, the response obtained is represented in Bode diagrams in which the impedance module and the phase angle serve as functions of the Log frequency, these diagrams indicate the sensitivity of the *EIS* technique to evaluate the presence of growth of a natural oxide on the steel surface, this is observed for the case of corrosion test in distilled water. Two well-defined time constants are observed in the evaluated frequency domain, one time constant at higher frequencies is related to the presence of an oxide layer, however, the intensity of the phase angle signal of 85° gives information about the oxide thickness and its adherence, however micro-cracks, closed porosity or growth defects are always present in many kinds of oxide layers that serve as conducting pathways of ions coming from the aqueous electrolyte, allowing electron charge

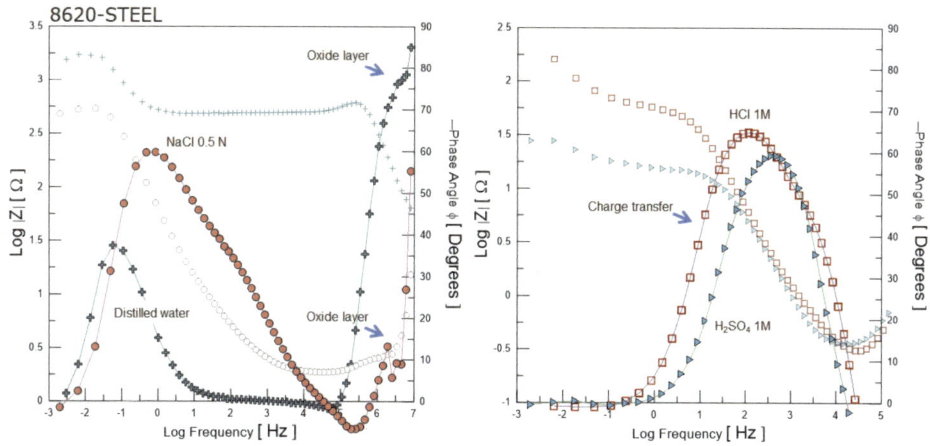

Figure 12.
Bode plots of impedance response of corroding 8620 plate at different aqueous solutions; distilled water, NaCl at 0.5 N, HCl at 1 M and H_2SO_4 at 1 M.

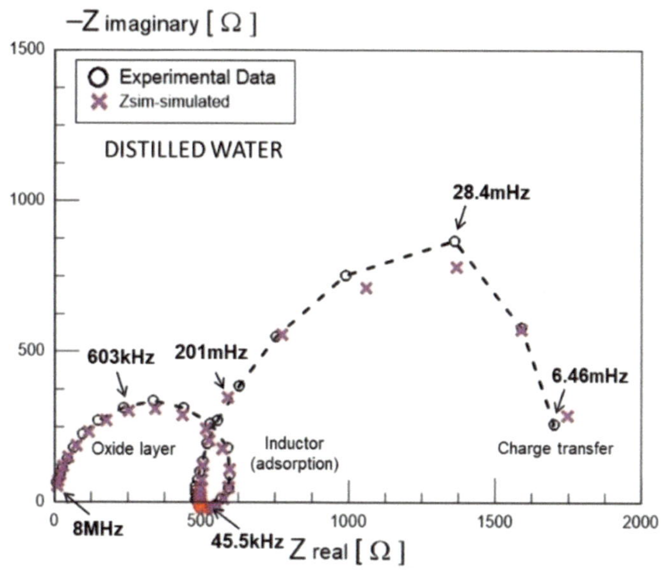

Figure 13.
Comparison of experimental and fitted EIS data for 8620 steel after exposure to distilled water.

transfer. This causes the phase-shifted continuously to zero degrees at frequencies between 80.7 kHz to 61.5 Hz, suggesting that the system behaves like a resistive component with a low flow of current near 10.3 $\mu A/cm^2$, *i.e.* the current signal oscillates with the same phase as the potential does. In this frequency range an adsorptive process is carried out in which ions passing through the oxide layer defects, this mechanism is shown by the inductive response of the **Figure 13**. However, at lower frequencies over 8.59 Hz, an increase in the phase angle to 40° (56.6 mHz) is observed as if it were a capacitor in which the steel interface is charged by OH^- molecules, it is worth mentioning that this response is not related to the corrosion process, but this is a typical response to a passive system with a magnitude of impedance about $10^3 \Omega$ -cm^2.

On the other hand, when the pH of the aqueous solution decreases to an acidified stage by the presence of ions such as Na^+, Cl^-, OH^-, SO_4^-, H^+, the shape of the impedance diagrams has been change, for example, for NaCl solution, a slightly acidified substance breaks-out almost the integrity of the natural oxide layer that covers the metal matrix and the response related to ion charge transfer to the metal interface is observed at lower frequencies. In addition to, an increase in current is also observed of about 34.479 μA/cm^2 and a |Z| of 10^2Ω -cm^2. Whereas, the same steel exposed to a more corrosive electrolyte such as HCl or H_2SO_4 at 1 M, the EIS response shows a single time constant that corresponding to the reaction's oxidation and reduction on the steel interface. That means, transient electrical charge events occur on the electrochemical double layer with ions, and is characterized by an increase of the current from 43.58 and 198.25 μA-cm^2, respectively and the decrease of one order of magnitude of the impedance module 10^1Ω -cm^2, that is, less resistivity. The results in **Table 3** indicates the simulation of impedance parameters with an appropriate electrical circuit that have been describe before, these data suggest that a higher current passing and large electrical charging at the interface of the steel increases the susceptible to attack by corrosion, that is, the internal energy of the aqueous solution has the ability to degrade freely the steel by pitting corrosion.

Same behavior was observed for impedance-monitored corrosion tests for a 316 stainless steel plate (18.24 wt.%Cr, 8.07 wt.%Ni, 1.76 wt.%Mn, 0.5 wt.%Si, 0.27 wt. %Mo as principal alloying elements) after exposure to different aqueous solutions such as distilled water, 0.5 N NaCl, 0.5 N KCL, 1 M HCl or 0.5 M H_2SO_4, **Table 4** shows the dissolution reaction. The impedance spectra that is shown in **Figure 14** indicates one of the advantages of the *EIS* technique to evaluate the performance of metal interface in full immersed to aggressiveness conditions of different electrolytes. In this sense the natural film of chromium oxide that protects stainless steel against corrosion is remarkable in distilled water by the presence of one time constant at higher frequencies with an impedance value near to 1 MΩ-cm^2.

Electrolyte	E_{corr} (mV)	I_{corr} (μA-cm^2)	C_{dl} (μF/cm^2)	R_{ct} (Ω-cm^2)
Distilled water (circuit#5)	-654	10.3	21.01	1841.8
Sodium chloride, NaCl 0.5N (circuit#3)	-580	34.5	183.8	695.7
Hydrochloric acid, HCl 1M (circuit#1)	-438	43.5	379.5	61.5
Sulphuric acid, H_2SO_4 1M (circuit#1)	-489	194.3	515.2	15.0

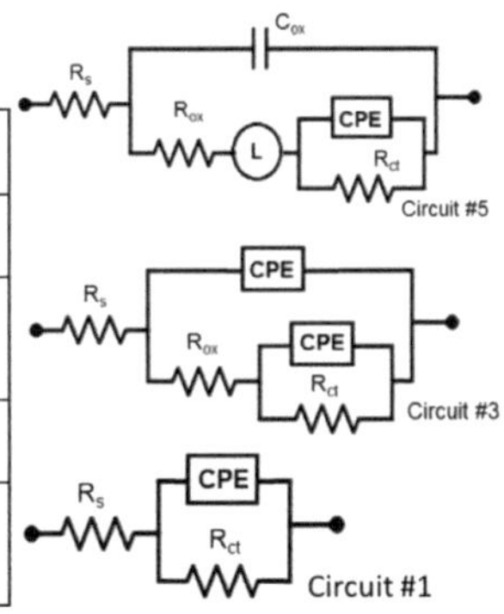

Table 3.
EIS parameters of simulated data to equivalent electrical circuit (EEC) *for the steel 8620 during its exposure to different electrolytes.*

Electrolyte	Concentration	C_{dl} (μF/cm^2)	Reaction
Distilled water	Pure-1lt	5.45	—
Sodium chloride, NaCl	29.2 gr/lt (0.5 N)	93.62	$NaCl + H_2O \rightarrow Na^+ + \mathbf{Cl}^- + OH^-$
Potassium chloride, KCl	37.27 gr/lt (0.5 N)	165.3	$KCl + H_2O \rightarrow K^+ + \mathbf{Cl}^- + OH^-$
Hydrochloric acid, HCl	15.56 ml/lt (1 M)	302.4	$HCl + H_2O \rightarrow \mathbf{Cl}^- + OH^- + H_{2\uparrow}$
Sulfuric acid, H_2SO_4	27.11 ml/lt (1 M)	313	$H_2SO_4 + H_2O \rightarrow \mathbf{SO_4^-} + OH^- + H_{2\uparrow}$

Table 4.
Capacitance of electrical double layer for the stainless steel SS316 during its exposure to different electrolytes.

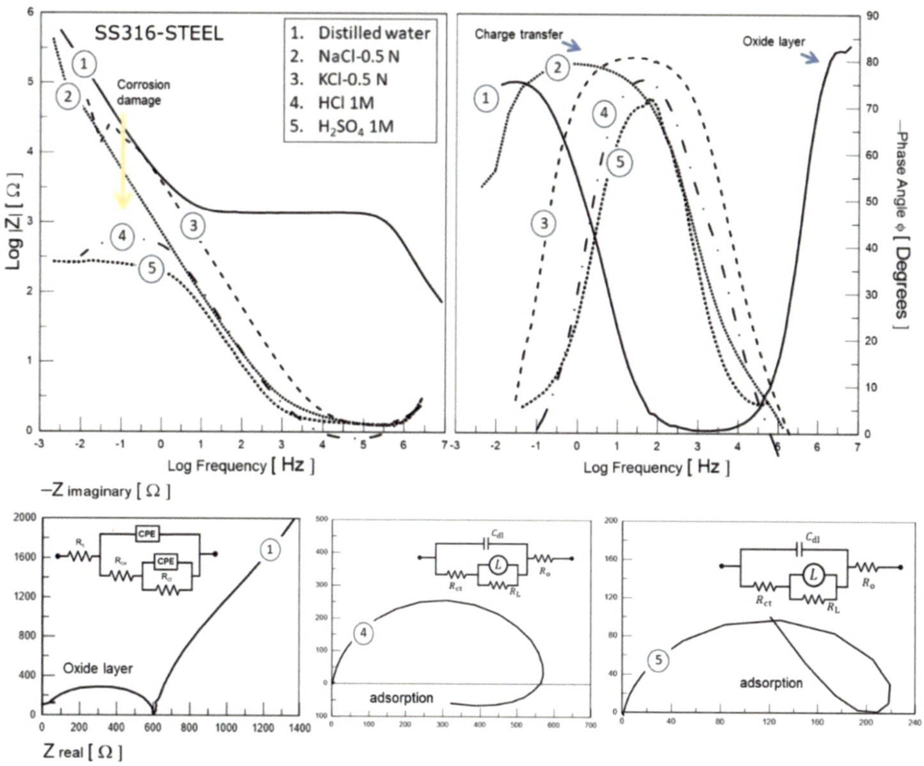

Figure 14.
Impedance response of corroding stainless steel (SS316) during exposure to (1) distilled water, (2) NaCl 0.5 N, (3) KCl 0.5 N, (4) HCl 1 M, and (5) H_2SO_4 1 M.

Meanwhile, the presence of Cl^- ions (NaCl or KCl salt) alters the coating interface, which is electrically charged by ions causing the passivity state of stainless steel broken-down due to the dissolution of the oxide film, it is assumed that the steel is susceptible to corrosion by pitting. This is also seen through the presence of a time constant in the frequency domain studied. Similarly, the experimental tests in stronger acid media (HCl or H_2SO_4) indicate that stainless steel is seriously corroded in these conditions as a decrease in the impedance value below 1 mΩ-cm^2.

2.3 Steel corroded at non-stationary solution (rotating disk electrode, RDE condition)

Other application of the *EIS* technique is like that shown in **Figure 15**, which is the evaluation of the effect on hydrodynamic conditions on the corrosion process in steels. This particular study has an interest to show the behavior of a pipeline steel (API-5 L-X70) that is used for transportation of hydrocarbon fluid. This steel was immersed in HCl 1 M solution at a different rotation speed of the working electrode (*WE*) from 0 to 1500 rpm, *i.e.* from static conditions 0 rpm, laminar flow 1 to 200 rpm and to turbulent flow 300 to 1500 rpm. **Figure 15** shows the *EIS* response in the representation of Bode and Nyquist for the steel interface during its exposure to a corrosive media at different flow rates.

At the steady-state conditions, without rotation, the impedance response is related to electrons flow from the aqueous media to the metal interface allowing the formation of an interfacial layer over the metal surface, called an electrical double layer or a thin oxide film, which is indicated by the distortion of the semicircle

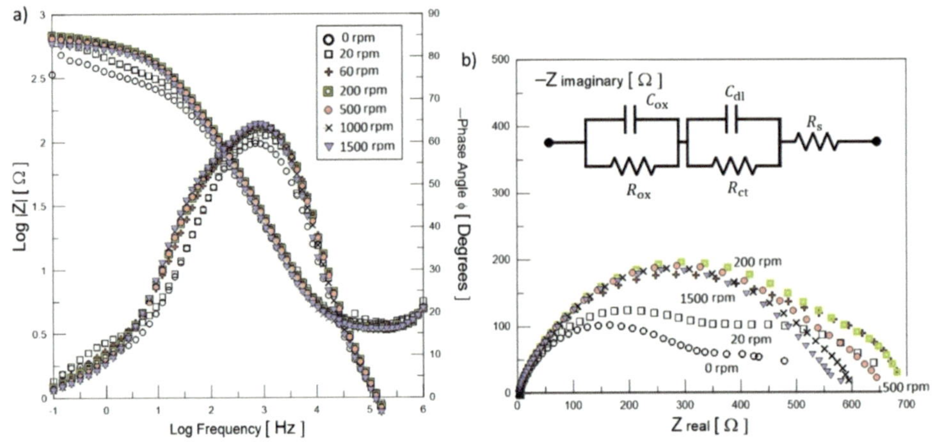

Figure 15.
Experimental impedance diagrams of corroding pipeline steel (API-5 L-X70) during exposure to HCl 1 M at different electrode rotation speed (0 to 1500 rpm). a) Bode plots representation and b) Nyquist complex plane.

presenting two time constants not very well-defined, besides in the diagram of bode two changes of slopes are shown for the impedance module. When applying rotation from 20 to 200 rpm an increase in the magnitude of the Z_{real} and Z_{imag} is observed due to the reaction kinetics at which the interfacial layer is forming at instantaneous rate and is controlled by electron charge and mass transfer mechanism. However, at turbulent conditions (>500 rpm) it does not allow the ions adsorption at the metal interface to maintain the presence of the double electrochemical layer or oxide film allowing only transients of electron transfer as a function of time, which promote the interfacial degradation of the steel. Therefore, the impedance diagrams show that under equilibrium conditions there is a corrosion rate controlled by the presence of a natural oxide on the steel surface, but this increased by the hydrodynamic conditions at turbulent flow, which is what is seen in real cases of application. But at moderate rotation speed the mass transport toward to the metal surface is carried out, giving opportunity to adsorption of molecules that come from the aqueous solution, which is consistent with the review literature [59].

2.4 Corrosion monitor in concrete reinforced materials

EIS technique can also be used for monitoring the evolution of the carbonation progress on concrete and the corrosion of the steel that serves as reinforcement. Carbonation results in a decrease in the pH of the cementation matrix when $CO_{2(g)}$ from the environment diffuses into the concrete structure, that can cause the loss of the passivity condition on the reinforcing steel surface and leads to an early failure of concrete by corrosion attack. Change in electrical resistance (R_{po}) and capacitance (C_{po}) of the concrete bulk is measured by a semicircle at high frequency region, which is the typical response of EIS diagram as that shown in **Figure 16**. More details are available in the research of H. Herrera in 2019 [6]. The corrosion test of this study was carried out on a fresh cross section of concrete sample after 7, 14, 21, 42, 61, 84, 106 and 120 days of artificially $CO_{2(g)}$ exposure periods (carbonation process). The characteristic impedance diagrams (*EIS*) of the concrete specimens after carbonation process at different ages of $CO_{2(g)}$ exposure during immersion in tap water are shown in **Figure 16**.

The *EIS* spectra is displayed in the Nyquist plots (Z_{real} vs. $Z_{imaginary}$), these results show a single capacitive well-defined semicircle at higher frequencies followed by a straight line for 7 to 84 days of carbonation, which indicates the

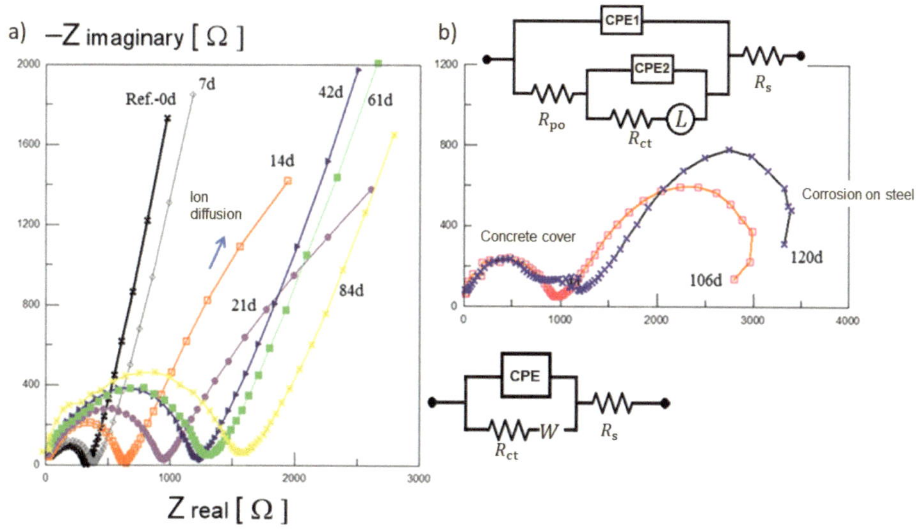

Figure 16.
EIS spectra for the particular system of concrete with reinforcing steel exposed to different days of a $CO_{2(g)}$ environment, carbonation process [6]. a) Nyquist complex plane showing the carbonation progress and b) Nyquist response for steel corroding.

specific resistance of the concrete that could be controlled by charge transfer process; while the straight line indicates a diffusion mechanism of ions through the pores. It is observed that the semicircle amplitude for the reference sample [REF.-0d, non-carbonated] is shorter than the carbonated samples at 7 or 120 days, this suggest that its resistance to the ions diffusion through the porous structure is much lower (a favorable condition for the ions coming from the aqueous solution driven easily into the porous structure of the concrete, resulting in the faster flow of electrons with chemical reactions and molecules adsorption processes around the vicinity of the steel interface), in addition to this, a typical signal describes a passive stage of the concrete. However, notable changes in the semicircle amplitude of the EIS spectra are observed, these changes are associated to the increase in electrical resistance (R) value of the concrete from 23.62 to 101.54 kΩ·cm^2 as the carbonation progress until to 84 days of $CO_{2(g)}$ exposure, this resulted to the blockade of the concrete pores by a calcium carbonate products, this reduces de alkalinity condition of the concrete matrix. However, the EIS diagrams for 106 days of exposure the resistance value decreases of about 58.26 kΩ·cm^2, the carbonation is almost complete, but after 120 days the resistivity still remains lower than 84 days of $CO_{2(g)}$ exposure (65.59 kΩ·cm^2) and the *EIS* spectra show remarkable changes in the low frequency domain. The changes registered by the *EIS* data for carbonated samples for 7 to 84 days are well-defined by one semicircle located at high frequencies (concrete porous resistance) with an infinite linear response at low frequencies (diffusion mechanism) only seen in the frequency domain of about $>10^6$ to 10^{-3} Hz by imposing a small amplitude of *AC* signal perturbation to the concrete/steel reinforcement system, this linear response was then modified by a second depressed semicircle with an inductive loop at lower frequencies in the domain of 10^{-6} Hz, using the EEC model #6 represents this behavior. The characteristic behavior of a second semicircle formed at lower frequencies for 106 or 120 days indicates that a process of corrosion may occur on the steel bar surface. The *EIS* parameters effectively demonstrate that after 106 days of exposure the carbonation is almost complete and corrosion damage is clearly progress on the steel bar. Carbonation progress was monitored by a significant increase in the diameter of the semicircle, thus demonstrating the increase in resistivity of ions transmission due to blockade of pores by precipitation of $CaCO_3$ compounds. Finally, the *EIS* technique results a practical tool for evaluating the carbonation progress on reinforced concrete structures without causing structural damage, and its sensitivity to predict the activation of the reinforcing steel to be corroded.

2.5 Inhibition in organic molecules or natural liquids extracted from plants

Particularly, the Mexican's oil-industry still uses tubular steel pipes for the specific purpose of transporting hydrocarbons or natural gas. Most of the lines are buried, so the national network extends over quite large distances, crossing varied terrains conditions some with rivers, others with salt-laden marshes, or polluted industrial or urban zones alike; the ambient temperatures and load pressure for the buried-pipelines network vary widely, to put it simply vulnerable to corrosion attack. Steel pipes are corroded as a result of iron oxidation during its exposure of longer service periods. Therefore, corrosion problems are directly related to ever-present economical and production losses, as well as environment affectations, though human losses also happen. Providing effective inhibiting substances that are added to processing fluids may reduce internal corrosion problems; there are a wide variety of organic substances known to act as corrosion inhibitors. **Figure 16** have demonstrated that small inhibitor quantities of organic molecules (2-Mercaptobenzimidazole, *MBI* or 5-Nitro-2-Mercaptobenzimidazole *NMBI* can be added to the media to diminish its inherent aggressiveness toward the steel surfaces [18, 60, 61]. It becomes evident that testing with the largest *2MBI* concentration, namely 200 ppm, there began to appear two-time constants, which suggests that two different processes are involved during the perturbation. One is related to a molecular adsorption mechanism of the organic compound over the polished metal surface, thus giving rise to multilayers, while the second constant is related to infiltration of the corrosive species through assorted passages formed during self-assembly and rearrangement of the organic molecules, very probably due to the diversity of interactive forces operating on the electrode system. This second time constant that operates at intermediate frequencies can be interpreted as a resistance to charge transfer. The *2MBI* inhibitor gave inhibiting efficiencies over 96% after adding only 20 ppm covering the metal surface exposed to the acid medium 1 M HCl. Therefore, the heterocyclic organic molecule *2MBI* was an efficient inhibitor in H_2SO_4 at 25 ppm. The plot of log Z vs. log f, shown in **Figure 17**, reveals that as the inhibitor concentration increases, so does the impedance, which is also related to the charge transfer resistance, R_{ct}. This value was obtained through fitting a *RC* electrical circuit model #3 to the experimental data. The $|Z|$ increment is explained by the excess of inhibitor's molecules in the solution, which as being bipolar it tends to adhere to the metal surface, also interacting among them thus forming a multilayered assembly, capable of blocking the electron charge transfer, refer to **Figure 17**, to appreciate more clearly the said $|Z|$ increase.

Furthermore, Natural liquid-extracts like *Morinda citrifolia* has been used as corrosion inhibitor for steels (AISI-1045) exposed to acidic environments of HCl. Both the organic and inorganic compounds commonly used in the industry to inhibit the corrosion process of metals and its alloys are mostly composed by highly toxic chemicals, in addition to being more expensive. In this research sugar-components derived from the *Morinda citrifolia* (MC) leaves have been extracted in aqueous solutions to perform a natural inhibitor capable to control de corrosion damage, which can replace the traditional inhibitors, being environmentally friendly [62, 63]. The experimental results indicate that this compound has shown excellent performance as corrosion inhibitor, reaching inhibition efficiency (*EI*),

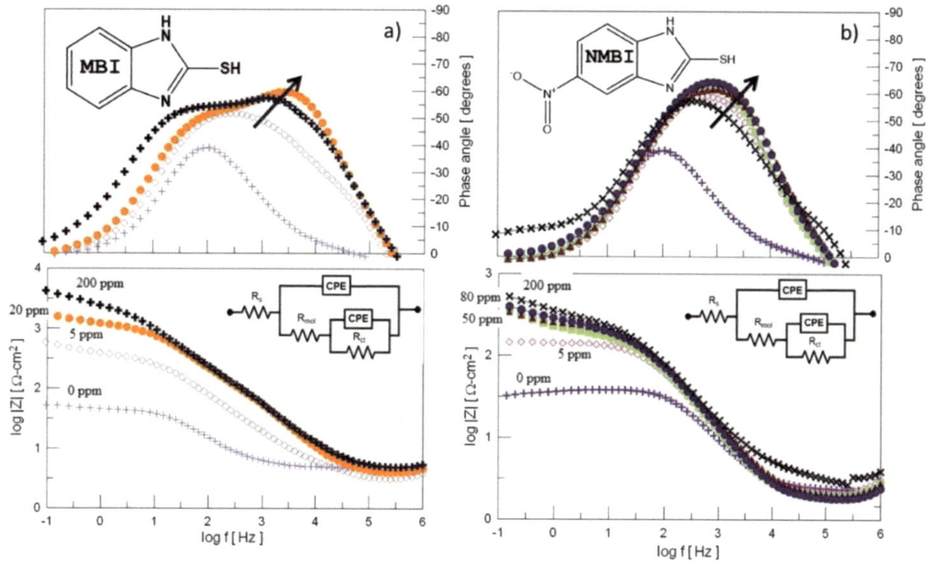

Figure 17.
EIS spectra in bode plots obtained from the pipeline steel API-5 L-X52 samples immersed in H_2SO_4 1 M as a function of the 2MBI or 5NMBI at different concentration [60, 61]. a) Response for 2- Mercaptobenzimidazole and b) response for 5-Nitro-2-Mercaptobenzimidazole.

values up to 90% at inhibitor concentrations ranging 0.8 to 2 g/L and immersion times of about 1 to 4 h. It has been found that the inhibition process takes place by the adsorption of the molecules on the surface of the metal (AISI 1045), by a physisorption mechanism. See **Figure 18.**

2.6 Hard-coatings as protection; borided treatment

Other attractive uses of the *EIS* technique are its application to evaluate the integrity and coating performance during its exposure in corrosive environments as a function of time. Actually, *EIS* is used as a quality control to evaluate the process

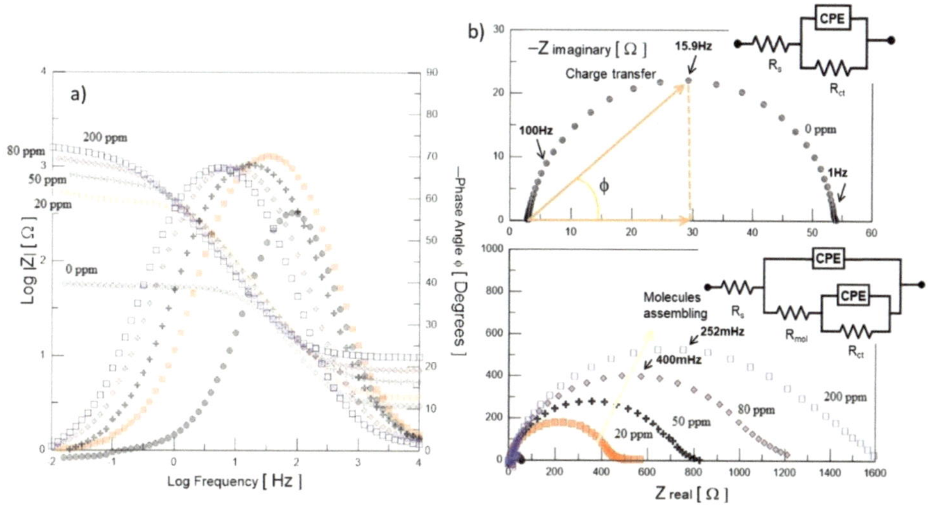

Figure 18.
EIS spectra in bode plots obtained from the pipeline steel API-5 L-X52 samples immersed in H_2SO_4 1 M as a function of the different concentration of natural molecules.

of surface finishing treatments in many industries. In this sense, the results of **Figure 19** show the characteristic impedance spectra that indicate the quality properties and corrosion resistance of a Fe_2B/FeB hard coating formed by boron atomic diffusion on the steel surface of a 1045 and 304 stainless steel during the boriding thermochemical treatment. Boriding is recognized as a thermochemical surface treatment in which boron diffuses into the ferrous substrate and reacts with Fe atoms of the bulk material to form a single (Fe_2B) or double-phase (Fe_2B/FeB) layer with a well-define thickness and composition [14]. The thickness of each layer has considerable effects on the mechanical behavior and corrosion behavior of the borided steels. However, the quality of the hard boride coatings depends essentially on the boriding temperature, treatment time, chemical composition of the steel substrate and the amount of boron atoms available around the sample surface to be coated.

In this study, in particular a powder-pack boriding was used on AISI-SAE 1045 steel and SS316 stainless steel as surface thermochemical treatment to improve hardness and wear resistance to the steel samples, due to its low cost of hard coating processing. Boriding can also enhance the corrosion resistance of ferrous materials as shown in **Figure 19**. The results indicate that a single boride layer of Fe_2B is

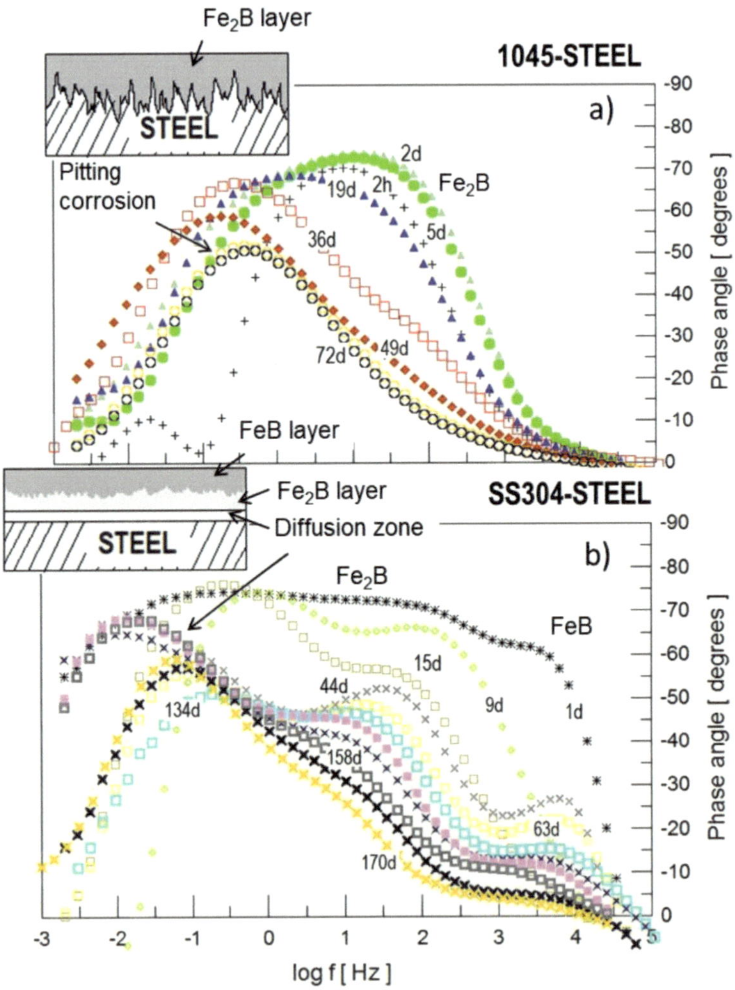

Figure 19.
EIS spectra for borided samples immersed in HCl 1 M as a function on exposure time. Boriding treatment was performed on AISI 1045 steel or AISI SS304 stainless steel treated at 950°C for 6 h [14]. a) Phase angle response for borided 1045 steel and b) Phase angle dependence for borided SS304 steel.

formed on the 1045 steel surface, its morphology consisting a deep saw-tooth derived from the existence of diffusion paths (porosity and micro-cracks) in the surface of the steel matrix, in which the boron atoms are interstitial inserted to the surface forming a stable phase. For the borided stainless steel SS304 at the same conditions forms two-well defined layers on the surface, the columnar phase that was growth on the 1045 steel is less intense for SS304, this is due to the high concentration of chromium and nickel on the substrate surface, so the diffusion of boron stops by reacting immediately to form interstitial compounds of CrB, Cr_2B or Ni_3B in combination with FeB and Fe_2B. EIS for the borided 1045 steel were recorded over 72 days of exposure to HCl 1 M solution, which the hard coating degrades slowly due to the defects on the coating structure that allow Cl^- ions infiltrate, this is denote by changing the *EIS* spectra shape from one time constant to two time constant with a clearly phase-angle shifted and loss of impedance value, that means pitting corrosion initiation. No-corrosion damage was observed for the borided SS304 during its exposure in HCl 1 M solution for at least 170 days. Three times constants were observed after 44 days that's reveal the presence of the FeB layer, after Fe_2B layer and the diffusion layer.

2.7 Steels used as beverages container

Steel-can containers are manufactured from thin metal plates and are commonly used for the distribution or storage of food or beverages. Most conventional steel beverage cans have bent to form a tube and then welding both sides leaving a firm seam, then joining the bottom end to the tube, finally, the steel can is filling-out with the content. However, it is necessary to mention that the steels cans have an internal polymer coating or have been treated by electroplating to coated internally with a thin layer of tin in order to prevent any oxidizing or electrochemical corrosion during the steel exposure to the liquid product that it contains, which could be carbonated soft drinks, alcoholic drinks, fruit juices, teas, herbal teas, energy drinks and others [64, 65]. Despite of this internal coating having the good quality, it may fracture during storage or dissolve in small amounts in the liquid product, which depends on certain factors such as temperature, stowage load and handling of the products during their storage, as well as the chemical composition of the liquid and steel. Due to this, efforts have been managed to replace tin-based coatings by chemical compounds derived from epoxy resins or polymers. Nevertheless, the set-up of the factors mention above may situate the metal container (e.g. steel cans) at a potential risk to develop internal corrosion.

On the other hand, the sale of beverages storage in steel cans are committed to their handling in warehouse, in this way, there is a predisposition of the people who buy drink-cans, they think if cans are struck or bent the coating has been damaged and could be associated that the liquid product is contaminated with Fe^+ ions. The impedance diagrams of **Figure 20** show that the *EIS* technique can be applied to assess the corrosion resistance of the internal coating in a specific beverage can.

In this case experimental corrosion tests on laboratory conditions were performed in a metal container used for the distribution of orange juice in Mexico. This can is made of steel with internally coated by a higher density polymer. Three particular cases are studied as denoted in the scheme of **Figure 20**; *EIS* spectra shown the behavior for a) with the coating, b) when the coating is mechanically damaged by a scratch and c) absence of coating, measured in HCl 1 M as a function of *AC* amplitude signal from 5 to 1000 mV. The bode diagrams indicate the presence of two well-defined time constants in the entire frequency domain for 5 and 10 mV of signal, the first one is related to the polymer coating with a resistance of electron -ion transfer of about $10^8 \Omega$ -cm^2 with a micro-porous net (conducting paths) inside the coating as indicated by the second time constant. However as increasing the amplitude of signal

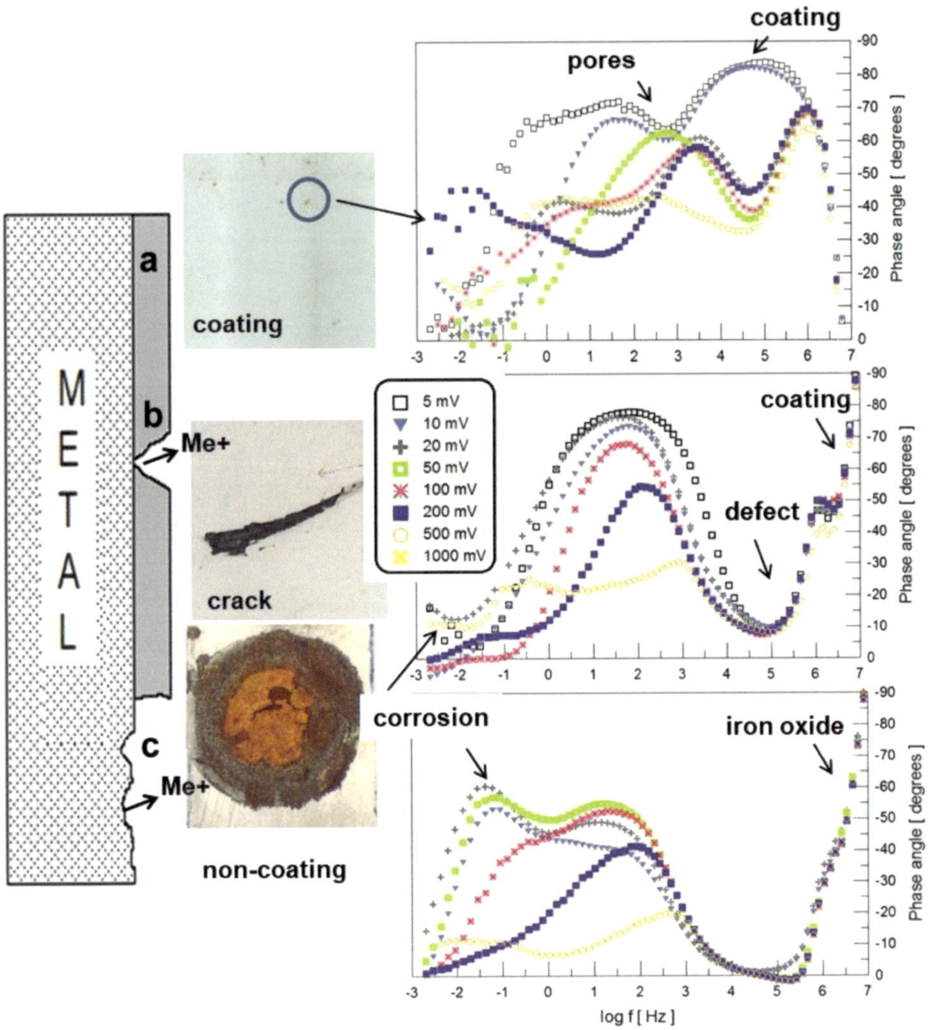

Figure 20.
Phase angle EIS response obtained for metal beverage containers at different surface condition after immersed in NaCl 0.5 M as a function on AC amplitude signal. a) Uniform polymer coating, b) scratch defect on coating, c) polished surface no-coating.

voltage the |*Z*| value drops below 10^4 Ω-cm^2, this response is associated with local stain-spots on the coating, which is indicted by a third time constant a low frequency. In the condition for the coating damaged by a localized defect such as a scratch or fracture, the impedance value decreases severely to 10^5 to 10^2 Ω-cm^2 as increased the *AC* signal, one time constant indicates the electron charge transfer processes through the defect that cause ions to be diffused below the coating until its failure. Finally, for the condition in the absence of the coating on the steel plate, the impedance diagrams show the corrosion process of the steel at different *AC* signal amplitudes, which shows severe corrosion after 200 mV showing 10^1 Ω-cm^2 of |*Z*| value.

3. Conclusions

This review study is related to the basic aspects of EIS to understand the corrosion mechanism of industrial steels that serve at different corrosive conditions, which has a great interest on giving an educational orientation and practical teaching guide of how to use the outstanding Electrochemical Impedance Spectroscopy (EIS) technique in metal corrosion technology. Therefore, this review considers a wide variety of practical electrochemical impedance cases based on the fundamental and qualities aspects of EIS theory and its experimental interpretation. This book chapter also serves as a support for postgraduate students to have a criterion in deciding through their own experiences when using the electrochemical impedance technique. The practical cases discussed here are part of the research experienced of Dr. Héctor Herrera Hernández (DR.3H) and his students & research group. It is worth to mention that EIS has been extended to various disciplines of science and technology, thus demonstrating great efficiency in evaluating the performance and integrity of metallic materials as can be seen in detail in the practical examples presented in this review work. So, EIS is not only applied to stationary conditions, but also more complex variables can be monitored such as: flow parameters, variable that undoubtedly represents the real conditions and could be an interesting challenge for analyzing and interpreting these phenomena by means of EIS data. The fitting EIS data using a mathematical model such as an equivalent electrical circuit is a critical process in the analysis and validation of EIS data for the acquisition of the system's electrical parameters that can be related to the corrosion rate of the material under study and also gives information of its capacity of electrons charge. Finally, EIS seeks to obtain information on the system and its evolution with time by applying a sinusoidal voltage as a function of frequency range, in order to determine the properties and feasibility of materials that serve under severe service conditions, such as industrial steels as is this case of the reviewed book chapter.

Acknowledgements

The authors dedicate this chapter to the memory of Professor Florian B. Mansfeld, USC.

The authors would like to acknowledge and express their gratitude to CONACyT for the SNI distinction as research membership and the monthly stipend received. Héctor Herrera Hernández (DR.3H) also would like to thanks to CIDETEQ and Secretaria de Investigación y Estudios Avanzados SIyEA/UAEM for their financial support through research project (4602/2018E). This project was conducted in the (Laboratory of Electrochemical and Corrosion of Industrial Materials at UAEM). Finally, DR.3H dedicates this work in memory to Professor **Florian B. Mansfeld**, for his teaching and guidance in the way of science (EIS technique), FBM will be remembered forever for his outstanding knowledge and contibutions.

Author details

Héctor Herrera Hernández[1*], Adriana M. Ruiz Reynoso[1],
Juan C. Trinidad González[1], Carlos O. González Morán[1],
José G. Miranda Hernández[1], Araceli Mandujano Ruiz[2], Jorge Morales Hernández[2]
and Ricardo Orozco Cruz[3]

1 Laboratorio de Investigación en Electroquímica y Corrosión de Materiales Industriales, Universidad Autónoma del Estado de México, Blvd. Universitario s/n, Predio San Javier Atizapán de Zaragoza, Estado de México 54500, México

2 Centro de Investigación y Desarrollo Tecnológico en Electroquímica, S.C., Parque Tecnológico Querétaro s/n, Sanfandila, Querétaro 76703, México

3 Instituto de Ingeniería, Universidad Veracruzana, S.S. Juan Pablo II s/n. Col., Zona Universitaria, Boca del Río, Veracruz 94294, México

*Address all correspondence to DR.3H: hherrerah@uaemex.mx

References

[1] Marcus P, Oudar J. Corrosion Mechanism in Theory and Practice. 1st ed. New York: Marcel Dekker, Inc.; 1995. 641p. DOI: 10.1201/9780203909188

[2] American Society for Metals. ASM Handbook, Vol. 13A Corrosion: Fundamentals, Testing and Protection. 10th ed. Ohio: ASM International; 2003. 2597p. DOI: 10.31399/asm.hb.v13b.9781627081832.

[3] Morcillo M, De la Fuente D, Díaz I, Cano H. Atmospheric corrosion of mild steel. Revista de Metalurgia. 2011;**47**(5): 1-19. DOI: 10.3989/revmetalm.1125

[4] Corrosion LD. Surface Chemistry of Metals. 1st ed. Vol. 614p. Florida: Taylor and Francis Group; 2007. DOI: 10.1201/9781439807880.

[5] Ahmad Z. Principles of Corrosion Engineering and Corrosion Control. 1st ed. Vol. 656p. Oxford: Butterworth-Heinemann; 2006. DOI: 10.1016/B978-0-7506-5924-6.X5000-4.

[6] Herrera Hernández H, González Díaz F, Fajardo SanMiguel G, Velázquez Altamirano JC, González Morán CO, Morales Hernández J. Electrochemical impedance spectroscopy as a practical tool for monitoring the carbonation process on reinforced concrete structures. Arabian Journal for Science and Engineering. 2019;**44**:10087-10103. DOI: 10.1007/s13369-019-04041-z

[7] Khatak HS, Raj B. Corrosion of Austenitic Stainless Steels. 1st ed. Alpha Science International: Pangbourne; 2002. 400p. DOI: 10.1002/maco.200490073

[8] Kelly RG, Scully JR, Shoesmith DW, Buchheit RG. Electrochemical Techniques in Corrosion Science and Engineering. 1st ed. New York: Marcel Dekker Incorporation; 2002. 440p. DOI: 10.1201/9780203909133

[9] Garfias-Garcia E, Colin-Paniagua FA, Herrera-Hernandez H, Juarez-Garcia JM, Palomar-Pardavé ME, Romero-Romo MA. Electrochemical and microscopy study of localized Corrosion on a sensitized stainless steel AISI 304. ECS Transactions. 2010;**29**(1):93-102

[10] Outokumpu. Handbook of stainless steel. 1st ed. Avesta: Outokumpu Oyj; 2013. 92p.

[11] Liang W. Surface modification of AISI 304 austenitic stainless steel by plasma nitriding. Applied Surface Science. 2003;**211**:308-314. DOI: 10.1016/S0169-4332(03)00260-5

[12] Liu RL, Yan MF. Improvement of wear and corrosion resistances of 17-4PH stainless steel by plasma nitrocarburizing. Materials and Design. 2010;**31**:2355-2359. DOI: 10.1016/j.matdes.2009.11.069

[13] Cardoso RP, Mafra M, Brunatto SF. Low-temperature thermochemical treatments of stainless steels. In: Mieno T, editor. Plasma Science and Technology: Progress in Physical States and Chemical Reactions. 1st ed. London: IntechOpen; 2016. pp. 107-130. DOI: 10.5772/61989

[14] Mejia-Caballero I, Martínez-Trinidad J, Palomar-Pardavé M, Romero-Romo M, Herrera-Hernández H, Herrera-Soria O, et al. Electrochemical evaluation of corrosion on borided and non-borided steels immersed in 1M HCl solution. Journal of Materials Engineering and Performance. 2014;**23**:2809-2818. DOI: 10.1007/s11665-014-1039-z

[15] Martini C, Palombarini G, Carbucicchio M. Mechanism of thermochemical growth of iron borides on iron. Journal of Materials Science. 2004;**39**:933-937. DOI: 10.1023/B:JMSC.0000012924.74578.87

[16] Natalya V. Likhanova NV, Domínguez-Aguilar MA, Olivares-Xometl O, Nava-Entzana N, Arce E, Dorantes H. The effect of ionic liquids with imidazolium and pyridinium cations on the corrosion inhibition of mild steel in acidic environment. Corrosion Science. 2010;52:2088–2097. DOI: 0.1016/j.corsci.2010.02.030.

[17] Goyal M, Kumar S, Bahadur I, Verma C, Ebenso EE. Organic corrosion inhibitors for industrial cleaning of ferrous and non-ferrous metals in acidic solutions: A review. Journal of Molecular Liquids. 2018;**256**:565-573. DOI: 10.1016/j.molliq.2018.02.045

[18] Abreu-Quijano M, Palomar-Pardavé M, Cuán A, Romero-Romo M, Negrón-Silva G, Álvarez-Bustamante R, et al. Quantum chemical study of 2-Mercaptoimidazole, 2-Mercaptobenzimidazole, 2-Mercapto-5-Methylbenzimidazole and 2-Mercapto-5-Nitrobenzimidazole as Corrosion inhibitors for steel. International Journal of Electrochemical Science. 2011;**6**: 3729-3742

[19] Krim O, Elidrissi A, Hammouti B, Ouslim A, Benkaddour M. Synthesis, characterization, and comparative study of pyridine relatives as corrosion inhibitors of mild steel in HCl medium. Chemical Engineering Communications. 2009;**196**(12):1536-1546. DOI: 10.1080/00986440903155451

[20] Palomar-Pardavé M, Romero-Romo M, Herrera-Hernández H, Abreu-Quijano MA, Likhanova NV, Uruchurtu J, et al. Influence of the alkyl chain length of 2amino 5 alkyl 1,3,4 thidiazole compounds on the corrosion inhibition of steel immersed in sulfuric acid solutions. Corrosion Science. 2012; **54**:231-243. DOI: 10.1016/j.corsci.2011.09.020

[21] Singh AK, Quraishi MA. The effect of some bis-thiadiazole derivatives on the corrosion of mild steel in hydrochloric acid. Corrosion Science. 2010;**52**(4):1373-1385. DOI: 10.1016/j.corsci.2010.01.007

[22] Bouklah M, Ouassini A, Hammouti B, Idrissi AE. Corrosion inhibition of steel in sulphuric acid by pyrrolidine derivatives. Applied Surface Science. 2006;**252**:2178-2185

[23] Ansari K, Yadav DK, Ebenso EE, Quraishi M. Novel and effective pyridyl substituted 1, 2, 4-triazole as corrosion inhibitor for mild steel in acid solution. International Journal of Electrochemical Science. 2012;**7**:4780-4799. DOI: : 10.1016/j.corsci.2011.09.020.

[24] Espinoza-Vázquez A, Negrón-Silva GE, González-Olvera R, Angeles-Beltrán D, Herrera-Hernández H, Romero-Romo M, et al. Mild steel corrosion inhibition in HCl by di-alkyl and di-1, 2, 3-triazole derivatives of uracil and thymine. Materials Chemistry and Physics. 2014;**145**: 407-417. DOI: 10.1016/j.matchemphys.2014.02.029

[25] Mansfeld F. Don't Be afraid of electrochemical techniques — But use them with care. Corrosion Science. 1988;**44**(12):856-868. DOI: 10.5006/1.3584957.

[26] Bard JA, Methods FRLE. Fundamentals and applications. John Wiley & Sons, Inc. 2nd. In: USA. 2001

[27] Mansfeld F, Lee CC, Zhang G. Comparison of electrochemical impedance and noise data in the frequency domain. Electrochimica Acta. 1998;**43**(3–4):435-438. DOI: 10.1016/s0013-4686(97)00060-1

[28] Mansfeld F, Lee CC, Zhang G. Comparison of electrochemical Impedance and noise data for polymer coated steel in the frequency domain. Materials Science Forum. 1998;**289-292**: 93-106. DOI: 10.4028/www.scientific.net/msf.289-292.93

[29] Mansfeld F. Use of electrochemical impedance spectroscopy for the study of corrosion protection by polymer coatings. Journal of Applied Electrochemistry. 1995;**25**(3):187-202. DOI: 10.1007/bf00262955

[30] Mansfeld F. Electrochemical impedance spectroscopy (EIS) as a new tool for investigating methods of corrosion protection. Electrochimica Acta. 1990;**35**(10):1533-1544. DOI: 10.1016/0013-4686(90)80007-b

[31] Mansfeld F, Xiao H, Wang Y. Evaluation of localized corrosion phenomena with electrochemical impedance spectroscopy (EIS) and electrochemical noise analysis (ANA). Materials and Corrosion. 1995; **46**(1):3-12. DOI: 10.1002/maco.19950460103

[32] Bolind AM, Lillard RS, Subbins JF. The electrical AC impedance response of oxide scales on 9Cr-1Mo steel immersed in molten lead-bismuth eutectic alloy at 200°C. Corrosion Science. 2015;**92**:48-57. DOI: :10.1016/j.corsci.2015.11.012.

[33] Shih H, Mansfeld F. A fitting procedure for impedance data of systems with very low corrosion rates. Corrosion Science. 1989;**29**(10): 1235-1240. DOI: 10.1016/0010-938X(89)90070-X

[34] Macdonald DD. Why electrochemical impedance spectroscopy is the ultimate tool in mechanistic analysis. ECS Transactions. 2009;**19**(20):55-79

[35] González-Morán CO, Miranda-Hernández JG, Flores Cuautle JJA, Suaste-Gómez E, Herrera-Hernández H. A PLZT novel sensor with Pt implanted for biomedical application: Cardiac micropulses detection on human skin. Advances in Materials Science and Engineering. 2017;**2017**:1-7. DOI: 10.1155/2017/2054940

[36] Macdonald DD. A method for estimating Impedance parameters for electrochemical systems that exhibit Pseudoinductance. Journal Electrochemical Society. 1978;**125**(12): 2062-2064

[37] Barsoukov E, Macdonald J.R. Impedance Spectroscopy theory, experiment, and applications. 2nd ed. Hoboken: A John Willey & Sons Incorporation; 2005. 595p.

[38] Lvovich VF. Impedance Spectroscopy: Applications to Electrochemical and Dielectric Phenomena. 1st ed. Hoboken: A John Willey & Sons Incorporation; 2012 353p

[39] Electrical Impedance CL. Principle, measurement, and aplications. 1st ed. broken Sound Parkway: Taylor & Francis. Group. 2013;**279p**

[40] Dalmont JP. Acoustic Impedance measurement, part I: A review. Journal of Sound and Vibration. 2001;**243**(3): 427-439. DOI: 10.1006/jsvi.2000.3428

[41] Alberts B, Johnson A, Lewis J, Raff M, Roberts K, Walter P. Molecular biology of the cell. 4th ed. Totnes: Garland. Science. 2002;**3786p**

[42] Gamry Instruments. Basics of Electrochemical Impedance Spectroscopy [Internet]. 2020. Available from: http:// https://www.gamry.com/application-notes/EIS/basics-of-electrochemical-impedance-spectroscopy/.html [Accessed: 2020-08-23]

[43] Perez N. Electrochemistry and Corrosion Science. 1st ed. Vol. 362p. New York: Kluwer Academic Publishers; 2004

[44] Mendoza-Flores J, Durán-Romero R, Genescá-Llongueras J. Espectroscopía de Impedancia Electroquímica en Corrosión. In: Llongueras JG, editor. Notas Técnicas Electroquímicas para el control y estudios de la corrosión. México: UNAM. 2002. pp. 55-91

[45] Mansfeld F. Shih h, Greene H, Tsai CH. Analysis of EIS data for common corrosion processes. In: Scully JR, Silverman DC, Kendig MW, editors. Electrochemical Impedance: Analysis and Interpretation. 1st ed. Fredericksburg: ASTM publication; 1993. pp. 37-53

[46] Mandujano-Ruiz A, Corona-Almazán E, Herrera-Hernández H, Morales-Hernández J. Opuntia ficus-indica (Nopal extract) as green inhibitor for Corrosion protection in industrial steels. In: Aliofkhazraei M, editor. Corrosion Inhibitors, Principles and Recent Applications. Croacia: Intech; 2018. pp. 145-158. DOI: 10.5772/intechopen.72944

[47] Hsu CH, Mansfeld F. Technical note: Concerning the conversion of the constant phase element parameter Y_0 into a capacitance. Corrosion. 2001; **57**(9):747-748. DOI: 10.5006/1.3280607

[48] Urquidi-Macdonald M, Real S, Macdonald DD. Applications of Kramers—Kronig transforms in the analysis of electrochemical impedance data—III. Stability and linearity. Electrochimica Acta. 1990;**35**(10): 1559-1566. DOI: 10.1016/0013-4686(90) 80010-l

[49] Matthew J, Orazem ME. On the application of the Kramers-Kronig relations to evaluate the consistency of electrochemical Impedance data. Journal Electrochemical Society. 1991; **138**:67-75

[50] Mansfeld F. Electrochemical impedance spectroscopy (EIS) as a new tool for investigating methods of corrosion protection. Electrochimica Acta. 1990;**35**(10):1533-1544. DOI: 10.1016/0013-4686(90)80007-b

[51] Mansfeld F, Kendig MW. Electrochemical Impedance spectroscopy of protective coatings. Materials and Corrosion. 1985;**36**(11): 473-483. DOI: 10.1002/maco.19850361102

[52] Mansfeld F, Tsai CH. Determination of coating deterioration with EIS: I. basic relationships. Corrosion. 1991;**47**(12): 958-963. DOI: 10.5006/1.3585209

[53] Tsai CH, Mansfeld F. Determination of coating deterioration with EIS: Part II. Development of a method for field testing of protective coatings. Corrosion. 1993;**49**(9):726-737. DOI: 10.5006/1.3316106

[54] Song YK, Mansfeld F. Development of a molybdate–phosphate–silane–silicate (MPSS) coating process for electrogalvanized steel. Corrosion Science. 2006;**48**(1):154-164. DOI: 10.1016/j.corsci.2004.11.028

[55] Amirudin A, Thierry D. Application of electrochemical impedance spectroscopy to study the degradation of polymer-coated metals. Progress in Organic Coatings. 1995;**26**:1-28. DOI: 10.1016/0300-9440(95)00581-1

[56] Yao Z, Jiang Z, Wang F. Study on corrosion resistance and roughness of micro-plasma oxidation ceramic coatings on Ti alloy by EIS technique. Electrochimica Acta. 2007;**52**(13): 4539-4546. DOI: 10.1016/j.electacta.2006.12.052

[57] Bosch RW, Moons F, Zheng JH, Bogaerts WF. Application of electrochemical Impedance spectroscopy for monitoring stress Corrosion cracking. Corrosion Science. 2001;**57**(6):532-539. DOI: 10.5006/1.3290379

[58] Gómez-García J, Rico A, Garrido-Maneiro MA, Múnez CJ, Poza P, Utrilla V. Correlation of mechanical properties and electrochemical impedance spectroscopy analysis of thermal coatings. Surface & Coatings Technology. 2009:812-815. DOI: 10.1016/j.surfcoat.2009.09.064

[59] Espinoza-Vázquez A, Negrón-Silva GE, Angeles-Beltrán D, Herrera-Hernández H, Romero-Romo M, Palomar-Pardavé MNEIS. Evaluation of pantoprazole as Corrosion inhibitor for mild steel immersed In HCl 1 M. effect of [pantoprazole], hydrodynamic conditions, temperature and immersion times. International Journal of Electrochemical Science. 2014;**9**: 493-509

[60] Espinoza A, Negrón G, Palomar-Pardavé ME, Romero-Romo MA, Rodríguez I, Herrera-Hernández H. Electrochemical Impedance spectroscopy analysis of 2-Mercaptobenzimidazole (2MBI) as Corrosion inhibitor in HCl 1M. ECS Transactions. 2009;**20**(1):543-553. DOI: 10.1149/1.326842.

[61] Cruz DY, Negrón G, Romero-Romo MA, Herrera-Hernández H, Palomar-Pardavé ME. Microwave-assisted preparation of 2-(Benzylthio)imidazole and 2-(Benzylthio)benzimidazole and its comparative Corrosion inhibiting performance with 2-Mercaptoimidazole and 2-Mercaptobenzimidazole. ECS Transactions. 2009;**20**(1):519-527. DOI: 10.1149/1.3268419.

[62] Barreto E. Extractos líquidos naturales de Morinda Citrofolia y Aloe-Vera como posibles inhibidores de la corrosión en aceros industriales [thesis]. Estado de México, Mexico: Universidad Autónoma del Estado de México; 2016

[63] Franco M, Herrera-Hernández H, García-Orozco I, Herrasti P. Extracto acuoso natural de Morinda Citrifolia como inhibidor de corrosión del acero AISI-1045 en ambientes ácidos de HCl. Revista de Metalurgia. 2018;**54**(4):1-10. DOI: 10.3989/revmetalm.128

[64] Rahayu EF, Asmorowati DS. Review of metal corrosion on food cans. Journal of Physics: Conference Series. 2019; **1321**:1-6. DOI: 10.1088/1742-6596/1321/2/022037

[65] Geueke B, Groh K, Muncke J. Food packaging in the circular economy: Overview of chemical safety aspects for commonly used materials. Journal of Cleaner Production. 2018;**193**:491-505

Chapter 2

Electrochemical Impedance Spectroscopy (EIS) Characterization of Kitchen Utensils Used as Materials for Local Cooking in Two Culinary Media

Jacques Sawadogo and Jean Boukari Legma

Abstract

This study is inscribed in the framework of the valorization of traditional kitchen utensils recycled from aluminum waste in Burkina Faso. In fact, these traditional kitchen utensils made of recycled aluminum alloys occupy a very important place in Burkina Faso's kitchen. The effect of foods for consumption on its local utensils was studied using the non-stationary technique and electrochemical impedance spectroscopy. For this purpose, a sample of utensil has been deducted on traditional production site. The corrosion behavior of the recycled aluminum alloy ok know chemical composition was evaluated by analyzing the impedance spectra obtained at the open circuit potential, in the salt media titrated at 3 $g{\cdot}L^{-1}$ and rice. Modeling electrical properties by using of a simple equivalent circuit made it possible to interpret the results obtained by impedance spectroscopy. The results showed a susceptibility to pitting corrosion and were confirmed by the electrochemical impedance spectroscopy method.

Keywords: corrosion, electrochemical impedance spectroscopy, equivalent circuit, foods, recycled aluminum alloy

1. Introduction

The pots made by craftsmen from recycled aluminum alloys play an important role in the cooking process in Burkina Faso. These alloys stand for very reactive materials and react instantly to media containing oxygen. This is why their outside surface is covered with an isolating oxide film. The thickness of this film reached around 10 nm and plays a protective role in those materials to corrosion which was generally observed in some aggressive media. The state in which these materials are located is called passivity state. The passivity condition can be stopped at any time when defects are found in the oxide film (discontinuity and heterogeneity) or the presence of aggressive ions in the electrolytic media (halogen, cyanide, etc.).

This can lead to a release of a localized aggression [1]. Aluminum alloys have a low density (2.7 g·cm^{-3}), a good thermal and electrical conductivity, a low melting point, easy to shape, a relatively low price, which is advantageous for local people [2]. Moreover, they are of high mechanical characteristics which make them to be utilized as structural materials. In Africa and particularly in Burkina Faso, craft industry turns to profit these properties in the recycling of aluminum alloys for kitchen utensils manufacturing; the raw material used in this field is made of combined or non-aluminum waste, from old car spare parts, beverage cans and tins [3]. Manufacture techniques remain empirical and recycled aluminum alloys are not homogeneous. Corrosion phenomena is favored when utensils are used for food cooking at high temperature or for long cooking time and when acidic or alkaline food are stocked in these same containers for long time [4, 5]. The humidity, the high temperature, and the cooking times are factors which favor the metallic materials corrosion, from which some of the component elements of the corroded material get through the surrounding aqueous media. Despite the numerous studies related to aluminum and their alloys corrosion, few scientific, strict and comprehensive studies on the behavior to corrosion, recycled allows in the craft industry have been conducted. The objective of this work was to study the corrosion behavior of a recycled aluminum alloy collected in the city of Ouagadougou (Burkina Faso) in various culinary media and to evaluate the anti-corrosion effect in these media. This study was carried out by the use of an electrochemical technique: electrochemical impedance spectroscopy (EIS).

2. Materials and methods

2.1 Sample preparation

The chemical composition of the recycled aluminum alloy is shown in **Table 1** [6]. Before each measurement, the aluminum allow surface preparation of these discs for electrochemical tests was the following: the discs were first ground with 400 through 4000 grit SiC papers and then polished with diamond down to 6 μm and followed by 1 μm alumina – 30% chrome oxide suspension, and finishing with 5% oxalic acid solution. Later, each polished sample was rinsed with acetone and put in an ultrasonic cleaner for 10 min. Subsequently, it was rinsed with milliQ water (with a conductivity of 0.7 μs/cm) and ethyl alcohol and finally dried under a hot air flow. The contact area with the culinary media is 3.46 cm^2.

2.2 Culinary media preparation

To simulate similar Burkina Faso operation, the testing media was local culinary media whose composition is given below. The media used in this study are: salt water (WS) titrated at 3 g·L^{-1} and broken rice (WR) in tap water (5 g of broken rice in 250 ml water) reserve for local consumption. The selection is made based on the fact that rice is the most consumed cereal in Burkina Faso. In this country,

Chemical elements	Al	Si	Fe	Cu	Zn	Mg	Cl	Mn	P	Pb	Remainder
Wt. %	82.8	12.76	0.76	1.21	1.27	0.48	0.19	0.27	0.22	0.01	0.03

Table 1.
Chemical composition of the recycled aluminum alloy determined by X-ray fluorescent technique.

the people consume on the average once daily prepared with vegetables, fish, and meats. These media were chosen to simulate a cooking process similar to that of Burkina Faso. All electrochemical measurements were performed in five replicates for each cooking media and show a reproducibility up to around 3–9%. Before each test, we assure that all the electrodes are submerged in the media, at the same depth in the electrochemical cell. As the cooking is most of the time performed at a hot temperature, the media were tested at boiling temperature (100°C) in order to simulate the real cooking conditions [7, 8].

2.3 Detection of aluminum in different culinary media using Eriochrome black T

2.3.1 Principle

Eriochrome Black T is a (3-hydroxy-4-[(1-hydroxy-2-naphthalenyl)azo]-7-nitro-1-naphthalenesulphonic) acid sodium salt, Mordant Black 11. In the presence of colored indicator [9], diluted aluminum in the buffer solution forms a complex which changes at wine-colored. The formed complex is more stable. Acidity of the obtained solution depended on aluminum content. The various measures of aluminum content in culinary media will be done by colorimetric with a spectrophotometer 7315 of JEWAY mark at 560 nm wavelength [10, 11].

2.3.2 Aluminum quantitative analysis method

Local kitchen utensil samples have been thoroughly washed and rinsed using distilled water then air dried. Each kitchen utensil has been filled out with studied solution WS and WR then heated at boiling temperatures. To compensate evaporation during heating phase, the final volume is adjusted to the end of each operation with distilled water [12].

2.3.3 Colorimetric dosage of aluminum released from local kitchen cooking

The loss of aluminum quantity released from two local kitchen utensils was determined by colorimetric dosage to 5 ml sample for each cooking solution. For that it was placed in the graduated flask of 50.0 ml containing 10.0 ml distilled water: 5.0 ml of EBT solution, 20.0 ml of buffer solution acetyl acetic acid ($C_4H_6O_3$) permit to hold a pH at 6; 1.0 ml ascorbic acid 2% and a volume of solution S_0 specified in **Table 2** then filled up to the line of gauge with distilled water. After stirring and resting during 20 minutes, samples were analyzed with spectrophotometer. The standard was measured with a solution without aluminum and tally with no absorbance. The concentration of aluminum in the different solution was expressed in mg/L.

2.4 Data analysis

Data obtained were analyzed for duration and temperature variations using the Student's t-test and XLSTAT 7.5.2 statistical software. Mean parameter concentrations were compared according to the Ryan-Einot-Gabriel-Welsch (REGWQ) test.

2.5 EIS-electrochemical impedance techniques

Electrochemical impedance spectroscopy (EIS) is a well-established quantitative method for the accelerated evaluation of the anti-corrosion performance of

	Control sample	S2	S3	S4	S5	S6	S7	S8	S9	S10	S11	S12
Volume of S0 (ml)	0	0.08	1.6	3.2	4	5	6.25	6.5	7.5	9	10	12.5
Distilled water	10	10	10	10	10	10	10	10	10	10	10	10
EBT (ml)	5	5	5	5	5	5	5	5	5	5	5	5
Buffer pH = 6 (ml)	15	15	15	15	15	15	15	15	15	15	15	15
Ascorbic acid 2% (ml)	1	1	1	1	1	1	1	1	1	1	1	1
Complete distilled water (ml)	50	50	50	50	50	50	50	50	50	50	50	50
Al concentration (mg/L)	0	0.016	0.032	0.64	0.8	0.1	1.25	1.3	1.5	1.8	2	2.5

Table 2.
Composition of standard scale.

protective coatings. Within short testing times, EIS measurements provide reliable data, allowing for the prediction of the long-term performance of the coatings. The result of EIS is the impedance of the electrochemical system as a function of frequency. EIS is a versatile testing procedure and can be performed under different conditions of stress, depending on the performance of the tested coatings. Electrochemical impedance spectroscopy (EIS) is a powerful technique that utilizes a small amplitude, alternating current (AC) signal to probe the impedance characteristics of a cell. The AC signal is scanned over a wide range of frequencies to generate an impedance spectrum for the electrochemical cell under test. EIS differs from direct current (DC) techniques in that it allows the study of capacitive, inductive, and diffusion processes taking place in the electrochemical cell.

2.6 Electrochemical measurements

The electrochemical measurements were conducted in the Analytical Chemistry and Interfacial Chemistry (CHANI) of the University of Brussels (ULB). The EIS measurements were determined in a three electrodes electrochemical cell containing the culinary media. There are three electrodes – the reference electrode, the auxiliary electrode, and the working electrode. A saturated calomel electrode (SCE) was used as the reference electrode, a platinum metal gate as the auxiliary electrode, and a recycled aluminum alloy as working electrode (WE) made in the laboratory. The EIS measurements were performed with employed Princeton Applied Research potentiostat (model PGSTAT 50). A microcomputer was used for data acquisition. The measurements were carried out after 60 minutes of cooking.

3. Results and discussion

3.1 Open-circuit potential measurements

Open-circuit potential, E_{OC}, changes were measured against a standard saturated calomel electrode placed in the same compartment. The recycled aluminum alloy was immersed in the culinary media exposing a circular area of about 3.46 cm^2. A copper wire was soldered at the rear of the electrode which was housed in a glass tube to protect it from the test culinary media. Results of the open circuit potential (E_{oc}) are shown in **Figure 1**. In the curve (**Figure 1**), a rapid increase of the open circuit potential was observed followed by a decrease of the value in the two culinary media. Open-circuit potential was studied for 15 min of cooking in the various culinary media. From the curve, a rapid increase of the open-circuit potentials followed by a decreasing of the value in the two culinary media (salt water at 3 $g.L^{-1}$ and broken rice) were observed. It can be noticed that these curves vary toward higher values during the first 150 seconds of cooking but after that, an almost decrease of the potential is observed. In this case, we can observe the aluminum passivation tendency which could have many forms: passivation caused by hydroxides which are absorbed at the metal surface, that caused by absorption of the existing components of the two cooking media or their combination.

A comparison of the behavior of recycled alloy in the media (broken rice and salt water) indicated that significantly higher corrosion potential was recorded in the salt water compared to broken rice media. This could be explained by their negative effect susceptible of influencing the passivation during the first minutes of cooking. According to literatures, the presence of chloride ions in study media could compete with media hydroxides ions when absorbed at the surface, allowing a localized corrosion and then a deterioration of the passive film [13]. In order to understand more about the existing behavior for metal/media in the cooking media, a series of curves was set out by electrochemical impedance spectroscopy in the context of comparative study in the different media.

3.2 Electrochemical impedance spectroscopy (EIS) measurements

Behavior to corrosion from recycled alloy in the two cooking media simulating a similar process to Burkina Faso cooking habit was studied by electrochemical impedance spectroscopy at 100°C and different cooking times. The frequency ranged from 100 KHz to 100 mHz, and the amplitude was set at 10 mV. Nyquist and Bode plots were used in broken rice media and that of salt water titrated at 3 $g.L^{-1}$ and up to boiling temperature after various cooking times in an open-circuit. Data

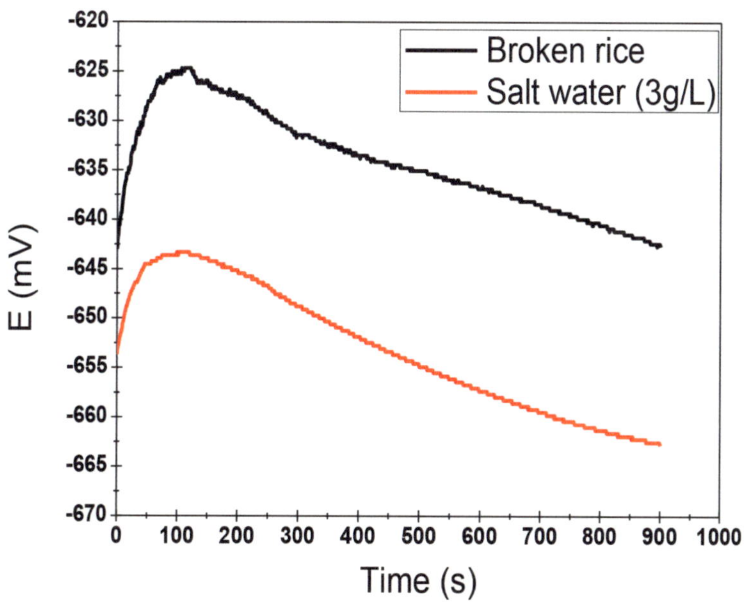

Figure 1.
Open-circuit potential for recycled aluminum alloy in salt water and broken rice.

acquisition and analysis were performed with microcomputer. The spectra were interpreted using the ZSimpWin program. These measures were performed in five replicates to ensure the results reproducibility.

3.3 Effect of cooking times

Measuring electrochemical impedance consists in studying the response of the electrochemical system, due to disturbance which is most often a low amplitude double signal. The strength of this technique is to differentiate the reaction phenomena from their relaxation times. Only quick processes are characterized in high frequencies; when the applied frequency decreases, appears the contribution of slower steps as transport phenomena or solution diffusion. To evaluate the behavior of the passive layer in various culinary media, the sample of aluminum alloy was immersed continuously for 60 minutes (00, 15, 30, and 60 minutes) for broken rice and salt water. During these cooking times, only measurements of impedances have been regularly performed since they do not disturb the system. Nyquist graph (**Figures 2** and **3**) illustrates the experimental impedance diagrams to corrosion potential obtained from the aluminum alloy in the studied culinary media. Indeed, **Figure 3** shows a progressive decrease in the size of the impedance spectrum in a more or less flattened half circle shape, characterizing the formation of the protective layer (alumina Al_2O_3). This leads to a decrease of the total recycled aluminum resistance with regards to the cooking time. In contrast to the salt water media, the broken rice media (**Figure 2**) show an increase in the spectra size, confirming the sample resistance of the media [14]. We find a phase difference with respect to axis of real (**Figures 2** and **3**), which may be explained by the surface none-homogeneity. However, for a better correlation between the experimental data and simulation, we introduced into the procedure for calculating a constant phase element and the surface none-homogeneity is realized through this constant phase element as follows (Eq. 1) [15–18].

Despite a constant phase element being utilized for data fitting instead of an ideal capacitor, since n values obtained from data fitting were in the range from 0.85 to 0.95, the value obtained from data fitting was taken as the capacitance (Eq. (1)).

$$Z_{CPE} = C\left[(j\omega)\right]^{-n} \tag{1}$$

Z_{CPE} = the impedance of the CPE;
CPE: constant phase element.
C = the capacitance associated to an ideal capacitor;
j = the complex imaginary number;
ω = the angular frequency and

$$-1 \langle n \langle 1$$

When n = 1, CPE is an ideal capacitor (Eq. (2)) [19, 20].

$$Z_{dc} = C\left[(j\omega)\right]^{-1} \tag{2}$$

Z_{dc} = double layer capacitance.

A true capacitive behavior is rarely obtained. The n values close to 1 represent the deviation from the ideal capacitive behavior [21].

The best simulation is obtained from the use of equivalent circuit proposed for metal/electrolyte interface and illustrated in **Figure 4**. This equivalent circuit was

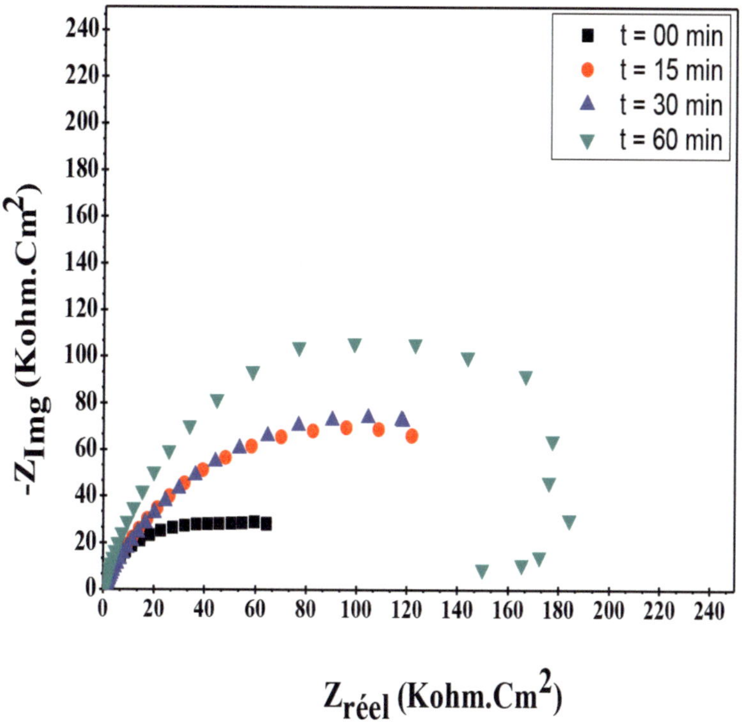

Figure 2.
Nyquist plots for recycled aluminum alloy tested in broken rice media.

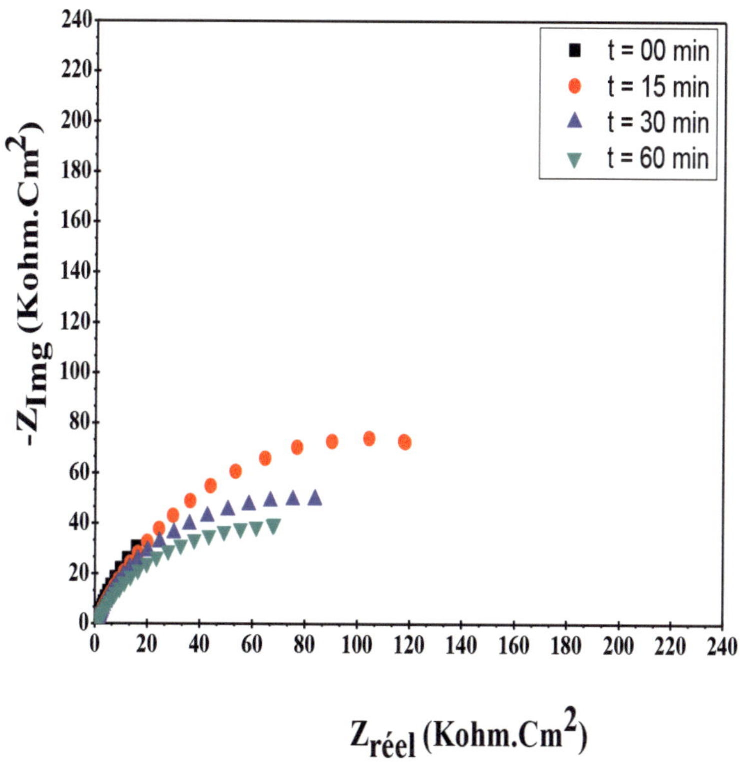

Figure 3.
Nyquist plots for recycled aluminum alloy tested in salt water media.

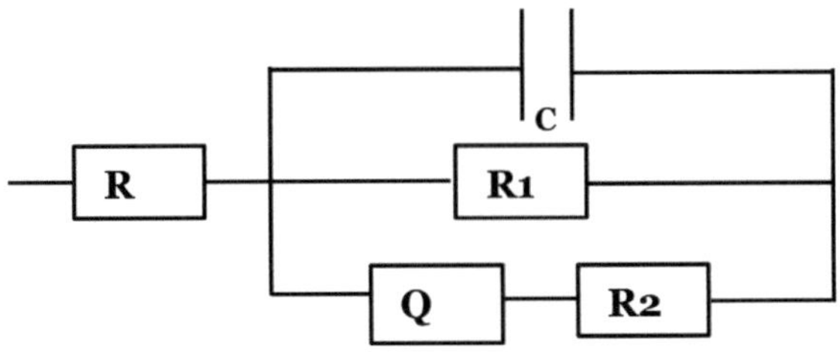

Figure 4.
Equivalent circuit to aluminum alloy in the cooking food of Burkina Faso.

Media	Times	R	C	R1	Q	n	R2
	(min)	kΩ·cm²	F·cm⁻²	kΩ·cm²	S sn·cm⁻²		kΩ·cm²
Broken rice	0	0.697	0.0017	0.3480	0.0048	0.73	75.4
	15	1.81.10-8	1.72.10-5	0.4830	0.0053	0.78	196.8
	30	3.21.10-8	1.53.10-5	0.5364	0.0060	0.76	223.8
	60	1.96.10-6	9.14.10-6	0.3061	0.0015	0.87	259.2
Salt water	0	6.70.10-11	1.41.10-5	0.4474	0.0426	0.80	330.9
	15	2.30.10-7	2.04.10-5	0.5586	0.0080	0.74	155.9
	30	2.40.10-8	1.53.10-5	0.5366	0.0060	0.76	222.3
	60	2.89.10-7	1.58.10-5	0.6573	0.0105	0.66	152.8

R, solution resistor; R1, polarization resistance; R2, oxide pore resistance; Q, constant phase; C, coating capacitance.

Table 3.
Electrical parameters of equivalent circuit obtained by fitting the experimental results of EIS tests.

proposed by Zhang et al. [9] to describe the bi-layer oxide film formed on aluminum and aluminum corrosion aqueous media.

This circuit is valid for all determinations. In the equivalent circuit, R is the salt water and the broken rice resistance, R_1 is the resistance to polarization, C is the corresponding capacity to the dense oxide layer, R_2 is the resistance in porous oxide position, and Q is the corresponding component of the constant phase to porous oxide positions. The results of the parameters in the equivalent circuit are shown in **Table 3**.

For the recycled aluminum alloy, different resistivity profiles in both media, regardless of the cooking time are observed as the impedance diagrams vary with the immersion time (**Figures 5** and **6**). It shows that parameters in the salt water media decrease in contrast to those in broken rice media for different cooking times up to 60 minutes. This behavior may be associated with physicochemical variations which occurred in the oxide film (alumina) during cooking in the salt water media (penetration of the electrolyte into the oxide layer and hydration of alumina) containing chloride ions. Comparison of the curves (**Figures 5** and **6**) clearly shows that resistivities in the layer alumina developed on the alloy recycled aluminum are higher in the broken rice media than those in the salt water. This could be explained by the presence of a more homogeneous and dense layer for the recycled aluminum in the media and also that of chloride ions in the salt water. Because, the behavior of interface/media is completely different with the latter. The overall behavior is reflected in the impedance diagram by a decrease in

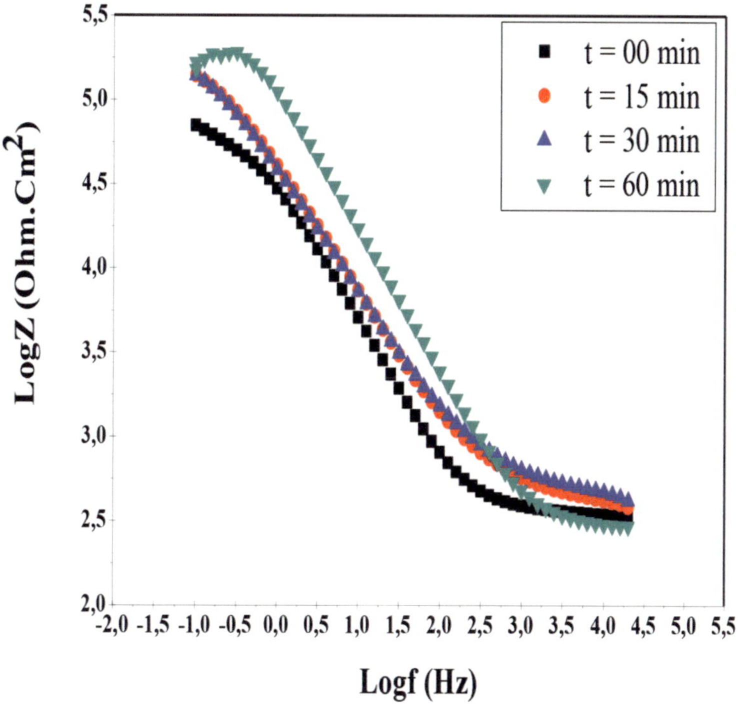

Figure 5.
Bode plots spectra for recycled aluminum alloy tested in broken rice media.

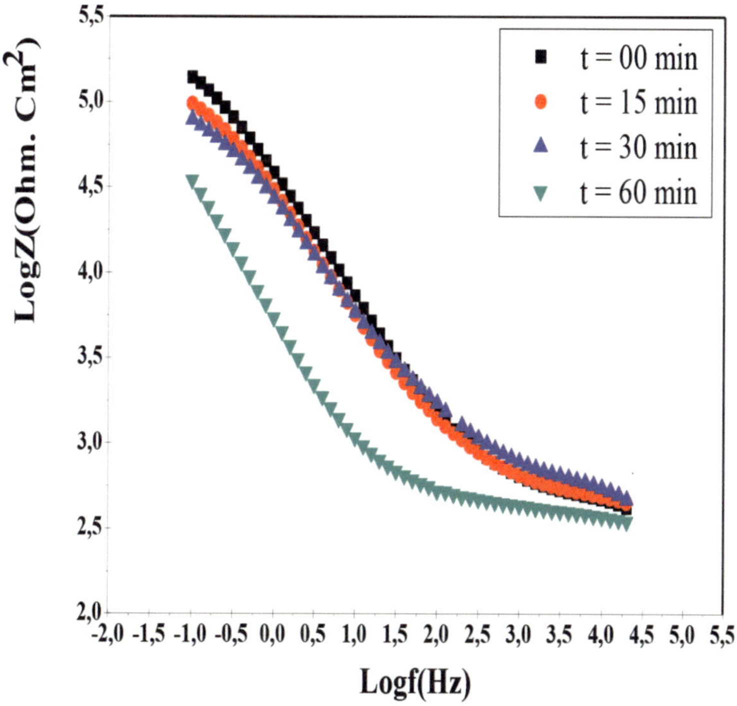

Figure 6.
Bode plots spectra for recycled aluminum alloy tested in salt water media.

size of the capacitive phenomenon. This can be explained by the weakening and destruction of a film which is likely to be developed on the surface of the studied alloy allowing disappearance of the distribution phenomenon and the decrease of the resistance. These differences may be explained by the oxide layer composition developed on the alloy which is influenced by the chemical composition of material solid media and by the chemical composition of the intermetallic particles [22–24].

In conclusion, resistivity profiles obtained for recycled aluminum alloy showed that the oxide layer developed is less protective in the salt water media than the broken rice. This result would be bound to the zinc presence which would return this less resistant system [25–27]. The negative effect of chlorides in the salt water media are presented in **Table 3**. This result was translated by the decrease in the polarization resistance. There also appeared an increase in the capacity associated with the polarization resistance. This increase may reflect the dissolution of the recycled alloy in the salt water media. The polarization resistance stands for the sum of the dense oxide layer resistance and that of the two cooking media (salt water and broken rice) [28]. In this case, R_2 is much larger than R_1, therefore, it can be considered as the polarization resistance. **Table 3** illustrated the simulation parameters. It shows that the polarization resistance increases gradually with the increased cooking time up to 60 minutes for media broken rice while for the salt water media, a decrease is observed followed by a slight increase. Highest values of the polarization resistance in broken rice media as compared to the salt water can be explained partly by the chemical composition of the recycled alloy capable of modifying the physical and chemical properties of the oxide layer into more or less noble depending on the studied media, and second, by the resistance of the charge transfer (R) which is not identical for both media. **Figure 7** indicated a clear difference between the polarization resistance values from the two cooking media.

Observation **Figure 7** curves show that the sample from the broken rice media is less corroded than that from the salt water media. This confirmed the destructive effect of the salt water media on our sample [29, 30].

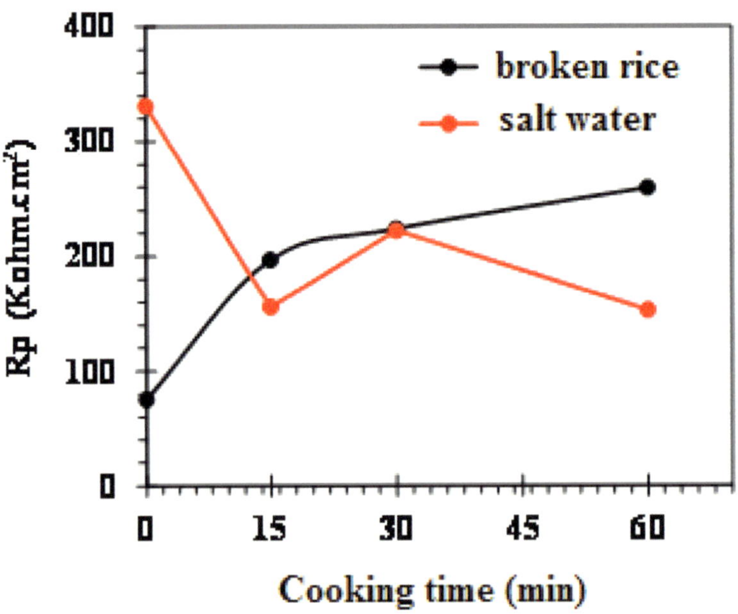

Figure 7.
Polarization resistance according to cooking times.

Sample	Times (min)	WS		WR	
D		25°C	100°C	25°C	100°C
	1440	5,84i		5.24j	
	15		57.85h		52.39g
	30		62.78e		56.82d
	60		70.78a		64.24a
ddl		3		3	
Probability		<0.0001		<0.0001	
Manning		HS		HS	

Results are means of 3 replications; HS = high significant. Test Ryan-Einot-Gabriel-Welsch (REGWQ), the difference is not significant between values added by the same letter in the same line.

Table 4.
Aluminum content measured in various cooking media of sample D after different contact times at room temperature (25°C) and at boiling temperature.

3.4 Influence of media, WS, and WR

Aluminum content released after 30 and 60 minutes in various cooking media is given in **Table 4.**

The same absorbance measured 30 and 60 minutes of cooking duration in the media WS (titrated at 3 $g.L^{-1}$ of salt) has given more important result in the other media (WR) of study. The high quantity of aluminum in this media has been probably linked to the presence of chloride ion and also to environmental pH. This result is according to the study conducted by Bommersbach and Duggan [31, 32]. Similar increase of aluminum loss with the increase of alcohol-free drinks acidity package in the aluminum bottle. These contents are very comparable with those got using tape water, concentrated tomato, and media WR in the same conditions. Other minerals in the tape water added to chloride ion have a significant influence on aluminum leaching with local kitchen utensils. For media WR important contents of aluminum had been lost in the cooking media after 30–60 minutes in the four local kitchen utensils. These results are similar to those decrypted by some authors [33, 34], in cooking breaking rice (WR) found to be not aggressive operation for sample containing more silicon. Studies showed that concentrated tomatoes caused more effect on cooking utensils [35]. Acidity of this product is so probability equivalent to those of fresh tomatoes, that is surely again a consequence of their origin and mode of production. Contributions of water at room temperature and tomato are so low that aluminum quantities swallowed and are relatively independent form the proportion of rice water. But, toxicity norm by some authors [26, 36] do state of acceptable daily dose to 1 mg by kilogram of body weight for human. This dose is a maximal tolerable quantity by human organism above which aluminum became toxic for him [22, 37]. This simplified outcome showed that we are far from the critical threshold for which human health is in danger. From this study, we can conclude that kitchen cooking utensils in Burkina Faso have not involved in particular toxicological danger.

4. Conclusions

This study contributed to the characterization by electrochemical impedance spectroscopy of the local kitchen utensils used for cooking. From this study, we conclude that variations of the impedance spectra in Niquist Z diagram based on the cooking time confirms the development of a protective oxide layer (alumina) of this alloys in electrochemical tests, resulting in an increase of the polarization resistance jointly with a decrease in the capacity of the double layer. Electrochemical tests showed a good efficacy of the sample in the broken rice media and having a good resistance to corrosion comparatively to salt water media. The low resistance to corrosion of sample in the salt water media is certainly caused by chloride ions. Susceptibility to corrosion by pitting has been confirmed by the method of electrochemical impedance spectroscopy.

Analysis of two local kitchen utensils with known composition and with various cooking media frequently used in Burkina Faso showed that aluminum content released increases with temperature influence, cooking time, and media. However, insignificant values of aluminum concentration released at room temperature in all solution are may be caused by the short stocking time, may be a decreasing of stocking temperature or another factor not deal with in this study. This study permits to update literature data and must support agribusiness and socio-economic interest of local kitchen utensils made in Burkina Faso according to the area. As precaution to take for limit risk of aluminum migration in foods:

- Avoid using spoil kitchen utensils, aluminum migrate more easily when kitchen utensils are worn;

- Avoid cooking or preserving food in kitchen utensil in aluminum. Food will absorb more aluminum if it is cooked or preserved in kitchen utensil (pan, leaf,...) made in this material;

- Avoid cooking vegetable or acid foods as tomatoes, citrus fruit in aluminum utensil, products absorb more easily this material.

Acknowledgements

The lead author is grateful to the Academy of Research and Higher Education (ARES) for providing funds for her PhD study. He is also grateful to Professor Jean Boukari Legma, the CHANI laboratory of Belgium and the staff of the Burkina Faso Institute of Environment and Agronomic Research (INERA) for the laboratory assistant who contributed a lot to the success of the study during data collection and analyses.

Author details

Jacques Sawadogo[1]* and Jean Boukari Legma[2]

1 National Centre for Scientific and Technological Research (CNRST), Ouagadougou, Burkina Faso

2 Faculty of Science and Technologies, Saint Thomas Aquinas University (USTA), Ouagadougou, Burkina Faso

*Address all correspondence to: jacques.sawadogo@ulb.ac.be

References

[1] Picot A. Intoxication de l'organisme par les métaux lourds et autres toxiques. Le mercure, le plomb et le cadmium trois métaux traces toxiques. 29 Novembre 2003 ; Conférence ADNO 2003, Paris; 2003. pp. 1-14

[2] DiBari GA, Read HJ. Electrochemical behavior of high purity aluminum in chloride containing solutions. Corrosion. 1971;**27**(11):483-494. DOI: 10.5006/0010-9312-27.11.483

[3] Sawadogo J, Bougouma M, Ramdé T, Boubié G, Bonou LD, Ogletree M-PD, et al. Caractérisations physico-chimiques des ustensiles de cuisine (marmites artisanales) fabriqués au Burkina Faso. Journal de la Societe Ouest Africaine de Chimie. 2014;**37**:18-28

[4] Draissia M, Debili M, Millet J-P. Comportement à la corrosion du système de films minces aluminium-cuivre. Journal of New Materials for Electrochemical Systems. 2005;**8**(3):229-233

[5] Fekrache A, Yacine Debili M, Lallouche S. Structural characterization of a bulk and nanostructured Al-Fe system. Materiali in tehnologije. 2014;**48**(5):631-637. DOI: 669.715:532.6:669.058

[6] Sawadogo J, Nisol B, Bougouma M, Ramdé T, Boubié G, Buess-Hermann C, et al. Caractérisations chimiques des ustensiles de cuisine artisanale en aluminium fabriqués au Burkina Faso: cas de Ouagadougou. International Journal of Biological and Chemical Sciences. 2014;**8**(6):2813-2827. DOI: 10.4314/ijbcs.v8i6.38

[7] Ndiaye MB, Bec S, Coquillet B, Cissé IK. Analysis of cooking utensils containing aluminum scrap in Senegal. Research Journal of Applied Sciences, Engineering and Technology. 2012;**4**(4):367-370

[8] Sawadogo J, Ayouba Mahamane A, Bambara D, Kabore A, Moutari SK, Some K, et al. Detection of aluminum in different culinary media using Eriochrome black T. International Journal of Biological and Chemical Sciences. 2017;**11**(6):3066-3074. DOI: 10.4314/ijbcs.v11i6.40

[9] Mahob RJ, Ndoumbe-Nkeng M, Ten Hoopen GM, Dibog L, Nyassé S, Rutherford M, et al. Pesticides use in cocoa sector in Cameroon: Characterization of supply source, nature of actives ingredients, fashion and reasons for their utilization. International Journal of Biological and Chemical Sciences. 2014;**8**(5):1976-1989. DOI: 10.4314/ijbcs.v8i5.3

[10] Jones L, Thurman D. The determination of aluminium in soil, ash and plant materials using Eriochrome cyanine RA. Plant and Soil. 1957;**9**(2):131-142

[11] Sangale M, Daptare S, Sonawane V. Determination of aluminum an magnesiumions in some commercial adsorptive anatacids by complexometric titrations. International Journal of Advanced Scientific and Technical Research. 2014;**4**(4):452-465. Available from: http://www.rspublication.com/ijst/index.html

[12] Sawadogo J. Caractérisation physico-chimique des ustensiles de cuisine fabriqués artisanalement au Burkina Faso [Thèse de Doctorat Unique]. Burkina Faso: Université de Ouagadougou; 2015

[13] Zhang F, Pan J, Lin C. Localized corrosion behavior of reinforcement steel in simulated concrete pore solution. Corrosion Science. 2009;**51**(9):2130-2138. DOI: i:10.1016/j.corsci.2009.05.044

[14] Yespica PJW. Etude comparative du comportement électrochimique des

alliages d'aluminium 2024 T351 et 7075 T7351 en milieu neutre de sulfate de sodium [Thèse de l'Institut National Polytechnique de Toulouse, France]; 2012

[15] Mohammad F, Al Zubaidy E, Bassioni G. Effect of aluminum leaching process of cooking wares on food. International Journal of Electrochemical Science. 2011;**6**:222-230

[16] Al Zubaidy EA, Mohammad FS, Bassioni G. Effect of pH, salinity and temperature on aluminum cookware leaching during food preparation. International Journal of Electrochemical Science. 2011;**6**(12):6424-6441

[17] Bougouma M, Sawadogo J, Kaboré A, Bambara D, Ayouba Mahamane A, Adamou R, et al. Analyse électrochimique du comportement des ustensiles de cuisine locaux dans l'eau de robinet et dans différents milieux culinaires. Afrique Science. 2018;**14**(1):106-118

[18] Karbouj R. Transfert d'aluminium: cas des matériaux pour contact alimentaire [Thèse de l'Institut National Polytechnique de Grenoble-INPG]. France, Paris; 2008

[19] Głuzek J, Masalski J, Furman P, Nitsch K. Structural and electrochemical examinations of PACVD TiO2 films in ringer solution. Biomaterials. 1997;**18**(11):789-794. DOI: 10.1016/S0142-9612(96)00210-4

[20] Hussin MH, Kassim MJ. The corrosion inhibition and adsorption behavior of Uncaria gambir extract on mild steel in 1 M HCl. Materials Chemistry and Physics. 2011;**125**(3):461-468. DOI: 10.1016/j.matchemphys.2010.10.032

[21] Gonzalez J, Mirza-Rosca J. Study of the corrosion behavior of titanium and some of its alloys for biomedical and dental implant applications. Journal of Electroanalytical Chemistry. 1999;**471**(2):109-115. DOI: 10.1016/s0022-0728%2899%2900260-0

[22] Ramdé T, Bonou L, Guel B, Legma JB. Comportement à la corrosion des alliages d'aluminium recyclés pour la confection de marmites. Journal de la Societe Ouest Africaine de Chimie. 2008;**26**:113-121

[23] Ramdé T. Comportement à la corrosion des alliages d'aluminium recyclés pour la confection de marmites [Thèse Unique de Université de Ouagadougou, Burkina Faso]; 2009

[24] Karbouj R, Desloges I, Nortier P. A simple pre-treatment of aluminium cookware to minimize aluminium transfer to food. Food and Chemical Toxicology. 2009;**47**(3):571-577. DOI: 10.1016/j.fct.2008.12.028

[25] Li JF, Ziqiao Z, Na J, Chengyu T. Localized corrosion mechanism of 2×××–series Al alloy containing S (Al2CuMg) and θ′(Al2Cu) precipitates in 4.0% NaCl solution at pH 6.1. Materials Chemistry and Physics. 2005;**91**(2):325-329. DOI: 10.1016/j.matchemphys.2004.11.034

[26] Chen X, Geng H, Li Y. Study on the eutectic modification level of Al–7Si alloy by computer aided recognition of thermal analysis cooling curves. Materials Science and Engineering A. 2006;**419**(1):283-289. DOI: 10.1016/j.msea.2005.12.036

[27] Snyder PO. The effect of aluminum and iron cooking utensils on food. Hospitality Institute of Technology and Management. 2008;**8**:1-2. Available from: http://www.hi-tm.com/

[28] Turhan S. Aluminium contents in baked meats wrapped in aluminium foil. Meat Science. 2006;**74**(4):644-647. DOI: 10.1016/j.meatsci.2006.03.031

[29] Kolics A, Polkinghorne J, Wieckowski A. Adsorption of sulfate and chloride ions on aluminum.

Electrochimica Acta. 1998;**43**(18):2605-2618. DOI: 10.1016/S0013-4686(97) 10188-8

[30] Ndiaye MB, Bec S, Coquillet B, Cissé IK. Analysis of the behavior of local cooking utensils in tap water, salt tap water and vegetable oil according to the temperature. Open Journal of Metal. 2013;**3**(4):80-85. DOI: 10.4236/ojmetal.2013.34012

[31] Duggan J, Dickeson J, Tynan P, Houghton A, Flynn J. Aluminium beverage cans as a dietary source of aluminium. The Medical Journal of Australia. 1992;**156**(9):604-605. Available from: https://pascalfrancis.inist.fr/vibad/index.php?action=getRecordDetail&idt=4382542

[32] Bommersbach P. Evolution des propriétés d'un film inhibiteur de corrosion sous l'influence de la température et des conditions hydrodynamiques [Thèse de Doctorat Unique]. Institut National des Sciences Appliquées de Lyon; 2005

[33] Wong KP, Alkire RC. Local chemistry and growth of single corrosion pits in Aluminum. Journal of the Electrochemical Society. 1990;**137**(10):3010-3015. DOI: 10.1149/1.2086150

[34] Sawadogo J, Bougouma M, Mahamane AA, Maman MA, Kabore A, Pouya MB, et al. La perte de masse dans divers milieux culinaires des ustensiles de cuisine locaux en aluminium: CAS du Burkina Faso. Afrique Science. 2016;**12**(3):160-170

[35] Zidane F, Rhazzar A, Blais J-F, Ayoubi K, Bensaid J, Basri SE, et al. Contribution à la depollution des eaux usées de textile par électrocoagulation et par adsorption sur des composés à base de fer et d'aluminium. International Journal of Biological and Chemical Sciences. 2011;**5**(4):1727-1745. DOI: 10.4314/ijbcs.v5i4.35

[36] Etteyeb N, Dhouibi L, Takenouti H, Alonso M, Triki E. Corrosion inhibition of carbon steel in alkaline chloride media by Na3PO4. Electrochimica Acta. 2007;**52**(27):7506-7512. DOI: 10.1016/j.electacta.2007.03.003

[37] Woodson G. An interesting case of osteomalacia due to antacid use associated with stainable bone aluminum in a patient with normal renal function. Bone. 1998;**22**(6):695-698. DOI: 10.1016/S8756-3282(98)00060-X

Chapter 3

Electrochemical, Thermodynamic, Surface, and Spectroscopic Study in Inhibition of Iron Corrosion with Turmeric Root Extract (TRE)

Khuloud Almzarzie, Ayman Almassri, Ahmad Falah and Hassan Kellawi

Abstract

Turmeric root extract was tested as corrosion inhibitor for iron in 0.5 M HCl, using potentiodynamic polarization and electrochemical impedance spectroscopy, scanning electron microscope, and energy dispersive X-ray analysis. The inhibition efficiency increases as the time of immersion rises but decreases with temperature rise. The Nyquist plots showed that the charge transfer resistance increases and the double-layer capacitance decreases as the time of immersion increases. Tafel results show that both corrosion current and corrosion speed are reduced with time of immersion. All impedance spectra of EIS tests exhibit one capacitive loop, which indicates that the corrosion reaction is controlled by charge transfer process. Inhibition efficiency increases with the concentration of the inhibitor reaching its maximum value, 88.90%, at 8 g/100 mL. Thermodynamic parameters, Ea, ΔH^*, and ΔS^*, were estimated, and the mechanism of corrosion and inhibition was discussed. The adsorption of turmeric root extract followed Langmuir adsorption isotherm.

Keywords: iron corrosion, electrochemical impedance spectroscopy (EIS), turmeric toot extract (TRE), double-layer capacitance (C_{dl}), scanning electron microscopy (SEM), inhibition efficiency IE (%)

1. Introduction

Inhibition of corrosion of iron is important for theoretical and practical aspects [1]. Iron and its alloys are of great importance in the industry, prompting vast research on corrosion resistance and its conducts [2]. Acids are wildly used in industries [3], such as industrial acid cleaning, acid descaling, and acid picking, and used to remove mill scale from metallic surfaces. Natural compounds containing sulfur, oxygen, and nitrogen atoms are effective as corrosion inhibitors in acid media; inhibitors are used to reduce the rate of dissolution of metals [4]. Natural plant inhibitors that contain organic compounds are widely used to inhibit minerals in an acidic medium [5] and are called environmentally friendly nontoxic inhibitors [6]. The adsorption of these compounds is influenced by the electronic structure of their content of active compounds, electrons density, aromatic rings, and functional groups possessing free electrons such as, R–OH –CHO, –N=N, etc. [7]. The adsorption of organic inhibitors at the metal/solution interface takes place through the replacement of water molecules by organic inhibitor molecules [8]. The efficiency of these compounds depends mainly on their abilities to be adsorbed on the metal surface with their polar groups moving as the reactive centers. The purpose of this research is to use a nontoxic environment friendly inhibitor formed from turmeric root extract to reduce iron corrosion and study its effect using electrochemical impedance and Tafel polarization methods. Turmeric root extract contains active substances represented by turmeric derivatives with a high percentage (90%), which are phenolic compounds of antioxidant. The most important are curcumin, demethoxycurcumin, and bisdemethoxycurcumin, which have the following chemical formulas (**Figures 1** and **2**) [9]:

Recent reference studies indicate the use of turmeric root extract in 2012 to inhibit copper metal. The inhibit corrosion was 98% with EIS technology and Tafel polarization [10]. In 2014, the effect of turmeric on steel was studied with Tafel polarization in a petroleum medium. The inhibit corrosion was 92% [11]. In 2017,

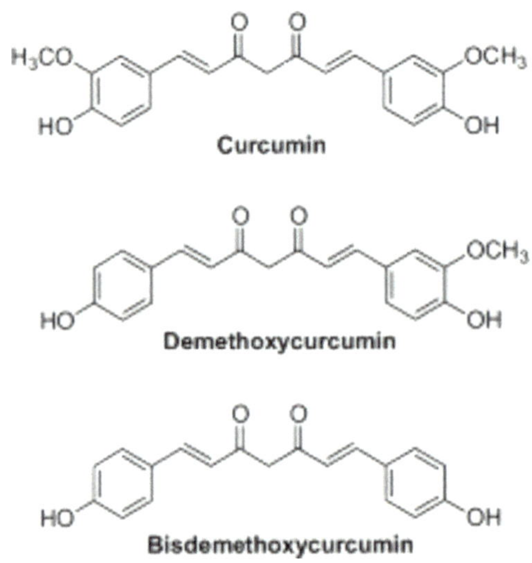

Figure 1.
Chemical formulas for turmeric derivative compounds [9].

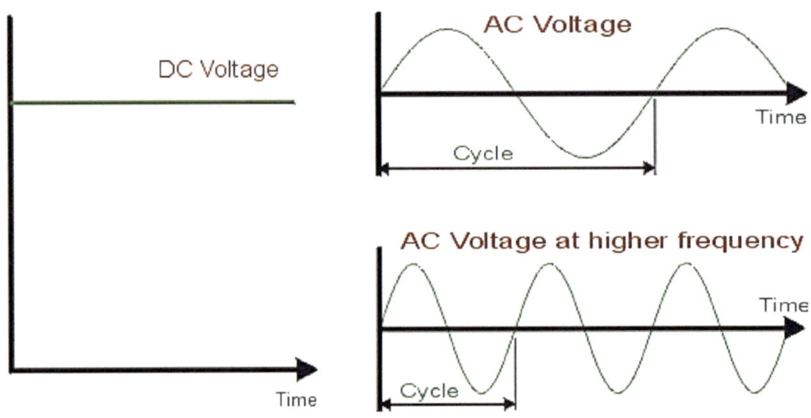

Figure 2.
Sinusoidal signal applied and the sinusoidal current signal produced by a specific phase difference.

the effect of turmeric on solid carbon in a saline medium (3.5% NaCl) was studied with Tafel technology; the inhibit corrosion was 91% [12].

This phenomenon has not been adequately studied in Syria; therefore, this research is an important step in studying the phenomenon of corrosion and understanding the mechanism of its occurrence and the conditions affecting it with the latest technologies Electrochemical impedance spectrometry and Mott-Schottky plot are used as a basis for future research in order to implement solutions on the ground.

1.1 Definition of electrochemical impedance spectrometry (EIS)

Electrochemical impedance spectrometry is also called alternating current (AC) and impedance spectroscopy (EI). It is a useful tool and the newest method in many studies, especially the study of corrosion of minerals. Impedance is defined as the measurement of the impedance of alternating current passage in each part of the electrical circuit, and it has units of ohm. Electromagnetic impedance spectroscopy measures the sinusoidal changes of the current and potential signals as a function of the frequency at the value of a current, a field of current, a value of a potential, or a field of potential. Frequency changes give rise to a different phase difference between the current and potential signals for each circuit element [1].

Impedance is measured by applying a sinusoidal signal or current and then measuring the resulting response to the corresponding variable in a given frequency range. The following equations show the signal of latency and current response at a specific frequency [3], which are found via Eq. (1):

$$E_t = E_0 \sin(\omega t) \tag{1}$$

where E_t is the latency at time t, E_0 is the applied signal amplitude, ω is the angular frequency (radians/second) equal to 2πf, and f (frequency) is the number of integrated vibrations per second (Hz/s).

The resulting current response is measured at the same frequency value with a phase difference between the current and potential signals and a difference in signal amplitude, which are found via Eq. (2):

$$I_t = I_0 \sin(\omega t + \varphi) \tag{2}$$

where φ is the phase difference in radians and I0 is the signal response amplitude. The relationship of impedance is given according to Ohm's law:

$$Z = E_t/I_t = E_0(\sin(\omega t)/I_0 \sin(\omega t + \varphi)) \tag{3}$$

$$= Z_0(\sin(\omega t)/\sin(\omega t + \varphi)) \tag{4}$$

where $Z_0 = E_0/I_0$.

The value of the phase difference at the value of a specific frequency is given by

$$\tan\varphi = Z_{img}/Z_{real} \tag{5}$$

where Z_{img} is the imaginary impedance and Z_{real} is the true impedance.

1.2 Features and applications

The electromagnetic impedance spectroscopy is distinguished as its results are accurate, as each point of the EIS curve contains information about the electrical process taking place, compared to conventional electrical methods that depend on measuring current changes, electrical charges, or the potential of the electrodes as a function of time, which depends on a certain value of its spectrum, such as a cyclic voltage, which includes current changes in hundreds of points for the potential to obtain a specific value for the peak oxidation and return at the corresponding current and potential value [3].

Among the applications of electromagnetic impedance spectroscopy, in addition to studying corrosion processes, electroplating processes, and semiconductors, it studies surface processes that include oxidation and reduction processes on the electrode surface, adsorption processes, electrical adsorption, and diffusion, as well as the kinetics of reactions in solutions, mass transfer, and resistance to solution, cells, and their electrical properties and batteries, determining the effect of each circuit element on impedance [1, 6]. It is also distinguished by not destroying the studied sample after testing [13].

2. Materials and methods

2.1 Preparation of plant extract

Turmeric root extract was prepared by washing and drying the turmeric root and grinding it then dissolving 1 g of powder in 100 mL methanol 50% and removing the solvent by placing the solution in a vacuum evaporator, at 60°C. Distilled water is used in the preparation process [14].

2.2 Preparation of metal specimen

The iron specimens have a composition (wt%) of 0.200% C, 0.500% Si, 1.600% Mn, 0.035% S, 0.035% P, 0.040% Nb, 0.012% N, 0.020% Al, and remaining Fe. This sample was analyzed in the Atomic Energy Commission and abraded with a series of emery papers 400, 1200, 1500, and 1800 grades. The samples were then washed thoroughly with distilled water and dried in air.

2.3 Test solution

A solution of 0.5 M concentrated acid was prepared using distilled water and 37% hydrochloric acid.

2.4 Electrochemical measurements

Electrochemical measurements were carried out using a potentiostat IVIUM-STAT.XR (Holland). A three-electrode cell system containing working electrode (iron coupon) of a 1 cm^2 exposed area, saturated (Ag/AgCl) electrode as a reference electrode, and then a platinum wire as auxiliary electrode was used. The electrochemical impedance spectroscopy measurements were carried out using the mentioned electrochemical system. Polarization curves were recorded at a sweep rate of 50 mV s^{-1}. Electrochemical impedance spectroscopy measurements were carried out at open-circuit potential over a frequency range of 1 MHz to 1 Hz. The sinusoidal perturbation has an amplitude of 0.01 mV.

2.5 Scanning electron microscopy (SEM) and energy dispersive X-rays (EDX)

The inhibitor film formation of the extracts surface was studied using SEM and EDX techniques.

3. Results and discussion

3.1 Mechanism of inhibition process

Turmeric root extract (TRE) used here as a corrosion inhibitor can serve as a scale inhibitor as well. This plant is characterized by the existence of a percentage of phenolic compounds (categories of curcumin) of a percentage up to 90%. It is a natural, nontoxic, environmentally friendly material. Active compounds in turmeric root extract are attributed to curcumin, demethoxycurcumin, and bisdemethoxycurcumin and to the multiple lone pair of electrons, multiple bonds, and/or conjugated л-type bond system [15]. Adsorption of these active molecules forms thin inhibitor films on the metal surface, which isolate the metal surface from the corrosive environment [16]. The oxygen atoms, the aromatic rings, and the bilateral bond of the aromatic rings boost the electronic pair freedom on the surface of the electrode. These compounds adsorb their free electrons on the surface of the electrode, and the iron is oxidized to form positively charged iron, thus forming a double electrical layer, and difference in voltage arises, as schematically presented in **Figure 3**. The inhibitor enhances the free electrons, which reduces iron corrosion and enhances inhibition.

3.2 Electrochemical impedance spectroscopy measurements

In the corrosion behavior of iron in 0.5 M HCl solution, in the absence and presence of TRE, it is investigated by the EIS, at 298 K after 1 hour of immersion in the acid solution. The double-layer capacitance (C_{dl}) and the frequency at which the imaginary component of the impedance is maximum ($-Z_{max}$) are found via Eq. (6):

$$C_{dl} = 1/w_{max}R_{ct} \text{ where } w_{max} = 2\pi f_{max} \tag{6}$$

The inhibition efficiency $\%IE_{Rct}$ that resulted from the charge transfer resistance (R_{ct}) is calculated by

$$\%IE_{Rct} = \left[\left(R_{ct} - R^0{}_{ct}\right)/R_{ct}\right] * 100 \tag{7}$$

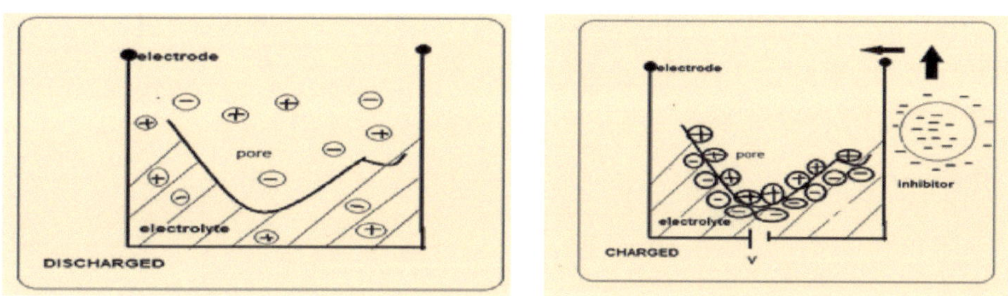

Figure 3.
Schematic presentation of the electric double-layer formation [Chem draw].

where R^0_{ct} and R_{ct} are the charge transfer resistance (R_{ct}) in the absence and presence of different concentrations of inhibitor, respectively [17].

Nyquist's and Bode's graphs of the results of the EIS of iron in 0.5 M HCl, in the absence and presence of different concentrations of TRE, were presented in **Figures 4** and **5**, respectively. The big capacitive loop refers to the adsorption of the inhibitor molecules (active compounds) presented in **Figure 2** on the iron samples [18]. R_s represents the resistance of the corrosive solution, and R_{ct} represents the effective resistance of the transport groups of the damper and the adsorption of their electrons onto the metal surface. C_{dl} is the amplitude of the formed double layer between the surface of the metal and solution.

Table 1 indicates that the R_{ct} values of the inhibited substrates increased with the concentration of inhibitors and C_{dl} values decreased because of the increased prevalence of active compounds from the inhibitor and the adsorption of their electrons on the iron surface, which confirms that TRE extract is an effective inhibitor of corrosion of iron in the medium of water chlorine acid [19]. The values of n between 0.7 and 0.98 indicate that the constant phase element Q operates as a capacitance in the equivalent electrical circuit, which indicates a complex. The adsorbent inhibitor is a capacitive capacitor in the electrical circuit equivalent to its positive bus, the surface of the solution, and the negative capacitor, the electrode surface [20]. Resistance values of the solution are small R_s due to the corrosion

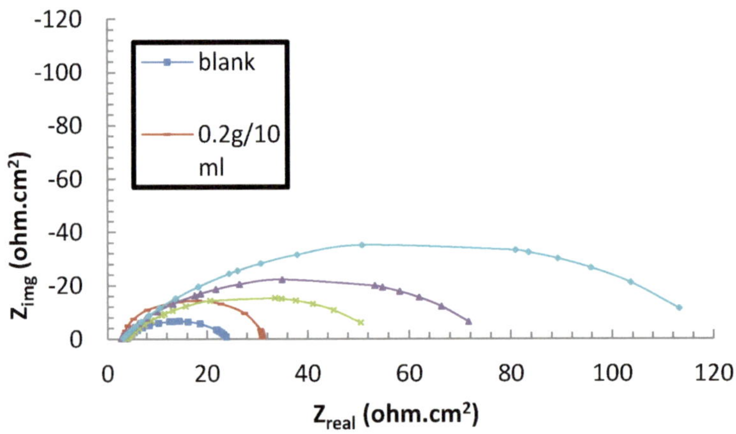

Figure 4.
Nyquist plots of iron in 0.5 M HCl containing varying concentrations of TRE after 1 hour of immersion in acid solution.

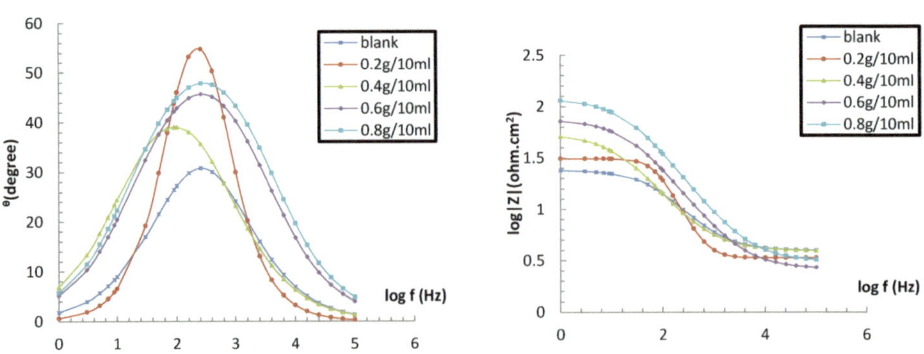

Figure 5.
Bode plots of iron in 0.5 M HCl containing varying concentrations of TRE after 1 hour of immersion in acid solution.

Inhibitor	Concentration (g/100 mL)	R_{ct} (ohm)	f_{max} (Hz)	C_{dl} (F/cm^2)	α	IE_{Rct}%
Blank	0.0	20.3	160	$4.90*10^{-5}$	0.73	—
Turmeric root extract	2.0	27.7	160	$3.59*10^{-5}$	0.98	26.71
	4.0	50.8	100	$3.13*10^{-5}$	0.70	60.04
	6.0	73.1	160	$1.36*10^{-5}$	0.70	72.23
	8.0	117	160	$8.51*10^{-6}$	0.70	82.65

Table 1.
Impedance parameters of corrosion of iron in 0.5 M HCl at 298 K in the absence and presence of different concentrations of TRE.

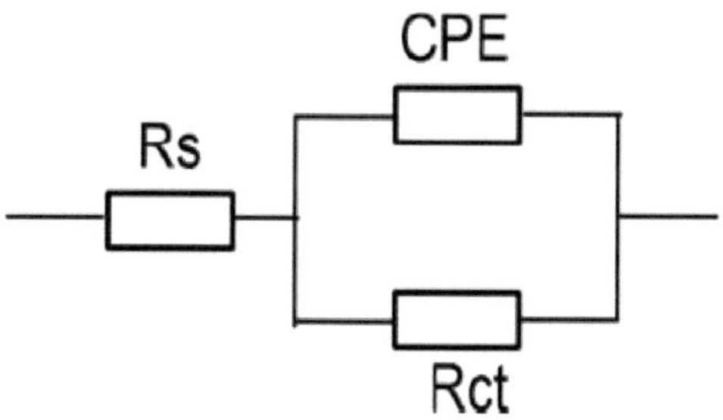

Figure 6.
Equivalent circuit model used to fit impedance spectra data.

Inhibitor	Concentration (g/100 mL)	E_{corr}- (mV/SCE)	I_{corr} (A*10^{-4}/cm^2)	βa (mV/dec)	βc (mV/dec)	CR (mm/year)	IE%
Blank	0.0	0.677	26.13	0.352	0.183	8.56	—
Turmeric root extract	2.0	0.3968	10.8	0.155	0.323	3.54	58.67
	4.0	0.474	4.7	0.146	0.233	1.55	82.01
	6.0	0.453	3.8	0.139	0.179	1.25	85.46
	8.0	0.417	3.4	0.131	0.225	1.11	86.99

Table 2.
Electrochemical parameters of iron in 0.5 M HCl solution at different concentrations without and with TRE.

concentration and temperature of the corrosive medium at different immersion times. Equivalent circuit model used to fit impedance spectra data is presented in **Figure 6**.

3.3 Potentiodynamic polarization measurements

3.3.1 Influence of concentration

Polarization measurements were done in order to know about the kinetics of the cathodic and anodic reactions. The anodic and cathodic current potential curves are extrapolated up to their intersection at a point where corrosion current density (I_{corr}) and corrosion potential (E_{corr}) are acquired [21]. **Table 2** shows the electrochemical parameters (I_{corr}, E_{corr}, βa, βc, and CR) obtained from Tafel plots for the iron electrode in 0.5 M HCl solution without and with various concentrations of

TRE. The I_{corr} values were used to calculate the inhibition efficiency, IE (%), in **Table 2**, using Eq. (8):

$$IE = [I - I_{corr}/I] * 100 \quad (8)$$

where I_{corr} and I are the corrosion current densities with the presence and absence of inhibitor, respectively. The CR values in **Table 2** used the following equation [22]:

$$CR = 3.27 * 10^{-3}\ i_{corr}\ E_w/d \quad (9)$$

where i_{corr} is the corrosion current density in micro A/cm^2, E_w is the equivalent weight of the corroding metal in grams, and d is the density of the corroding metal in g/cm^3.

Under the experimental conditions performed, the cathodic section of the plot represents the hydrogen evolution reaction, while the anodic section represents the iron dissolution reaction. They are determined by the extrapolation of Tafel lines to the respective corrosion potentials.

The results in **Table 3** indicates that the inhibitor reduces the corrosion current value and inhibition; IE (%) increases with the concentration of the inhibitor reaching 88. 90%, at 8 g/100 mL. This result suggests that good inhibition act for TRE [23].

T (K)	C (g/100 mL)	E_{corr}- (mv/SCE)	I_{corr} (μA/cm^2)	βa (mV/dec)	βc (mV/dec)	CR mm/year	IE%
283	Blank	0.677	26.13	0.352	0.183	8.56	—
	2	0.3968	10.8	0.155	0.323	3.54	58.67
	4	0.468	5.9	0.148	0.177	1.933	77.42
	6	0.453	3.8	0.139	0.179	1.25	85.46
	8	0.442	2.9	0.143	0.188	0.979	88.90
293	Blank	0.661	31.1	0.316	0.174	10.19	—
	2	0.379	8.35	0.142	0.328	2.73	73.15
	4	0.458	6.6	0.138	0.205	2.167	78.78
	6	0.458	4.8	0.164	0.206	1.581	84.57
	8	0.441	4	0.131	0.193	1.318	87.14
303	Blank	0.677	33.8	0.331	0.168	11.07	—
	2	0.452	13.5	0.167	0.249	4.4	60.06
	4	0.426	9	0.132	0.263	2.97	73.37
	6	0.474	7	0.152	0.193	2.29	79.29
	8	0.498	5.7	0.14	0.16	1.87	83.14
313	Blank	0.714	36.2	0.378	0.119	11.84	—
	2	0.477	14.6	0.188	0.237	4.788	59.67
	4	0.419	11	0.135	0.28	3.631	69.61
	6	0.379	7.9	0.133	0.339	2.6	78.18
	8	0.468	5.9	0.148	0.177	1.933	83.70

Table 3.
Polarization parameters of iron in 0.5 M HCl at different temperatures with various concentrations of TRE.

3.3.2 Influence of temperature

Polarization curves for the iron in 0.5 M HCl solution are shown in **Figures** 7 and 8 in two different conditions, with constant concentrations of TRE and in the presence of different concentrations of TRE in the temperature range 283–313 K. The surface overage (Θ) was calculated using

$$\Theta = \text{IE}\ (\%)/100 \tag{10}$$

The inhibition efficiency IE (%) is given by Eq. (8).

The results of the **Table 2** refer that temperature increase leads to I_{corr} increase, while the addition of TRE resulted in the decrease of the I_{corr} values across the temperature range. The results also indicate that the inhibition efficiencies increased with the concentration of inhibitor but decreased proportionally with temperature. Such behavior can be rationalized that the inhibitor acts by adsorption onto the metal surface [24].

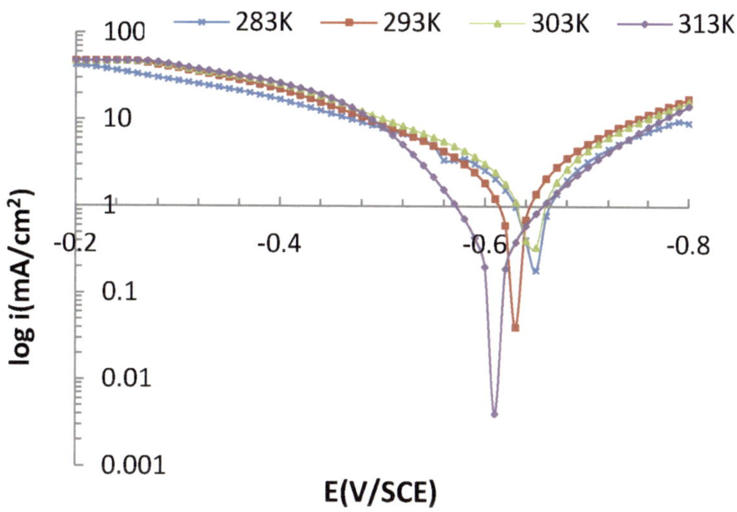

Figure 7.
Polarization curves of iron in 0.5 M HCl at different temperatures.

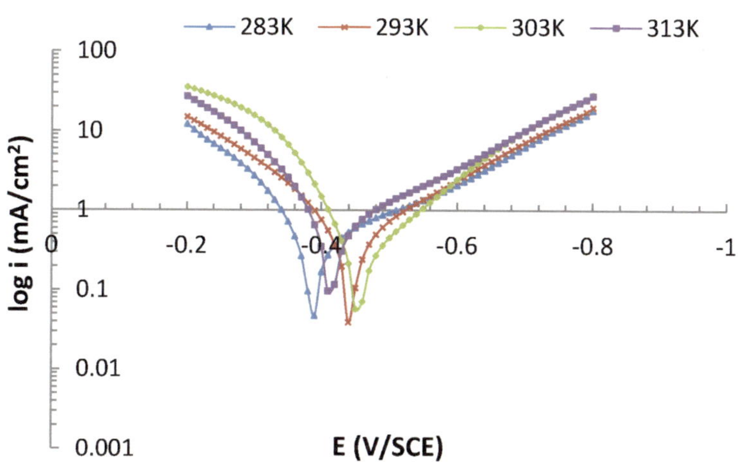

Figure 8.
Polarization curves of iron in 0.5 M HCl at different temperatures in the presence of 8 g/100 mL of TRE.

Ea, ΔS*, and ΔH*, for both corrosion inhibition and corrosion of iron in 0.5 M HCl in the presence and absence of TRE at different concentrations between 283 and 313 K, were calculated from an Arrhenius-type plot (Eqs. (11) and (12)) [25]:

$$\text{Log}\,(I_{corr}) = -E_a/2.303RT \tag{11}$$

where I_{corr} is the corrosion current density (taken from averaged polarization), Ea is the activation of ion energy, and R is the universal gas constant.

$$I = \frac{RT}{Nh}\exp\left(\frac{\Delta S*}{R}\right)\exp\left(-\frac{\Delta H*}{RT}\right) \tag{12}$$

where h is Plank's constant, N is Avogadro's number, ΔH^*_a is the enthalpy of activation, and ΔS*a is the entropy of activation.

The plots of Log (I_{corr}) vs. 1/T and Log (I_{corr}/T) vs. 1/T gave straight lines with the slope of -E_a/R and -ΔH* /R, respectively. The intercepts were A and [Ln (R/Nh) + (ΔS*/R)] for the Arrhenius and transition state equations, respectively (**Figures 9** and **10**). The calculated values of the activation energy Ea, the entropy of activation ΔS^*, and the enthalpy of activation ΔH^* are presented in **Table 4**.

The activation energy values with TRE film are greater than the absence as shown in **Table 4**. This result is consistent with previous results, since adsorption of active compounds on the metal surface impeded the corrosion reaction as if an additional energy barrier had arisen to dampen the corrosion reaction. The positive signal for a change in the enthalpy of the corrosion reaction indicates that the

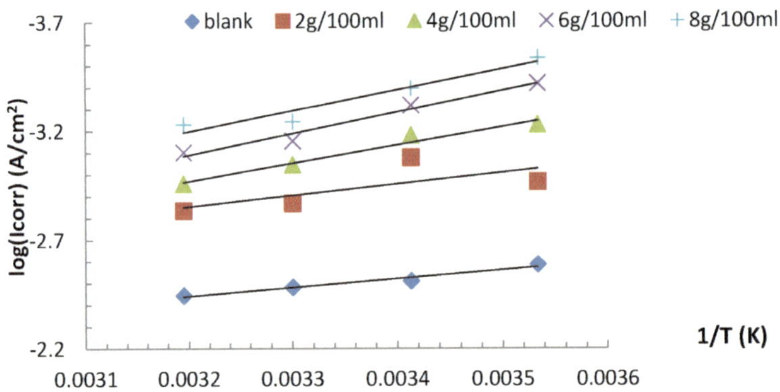

Figure 9.
Arrhenius plots of log (I_{corr}) versus 1/T at various concentrations of TRE.

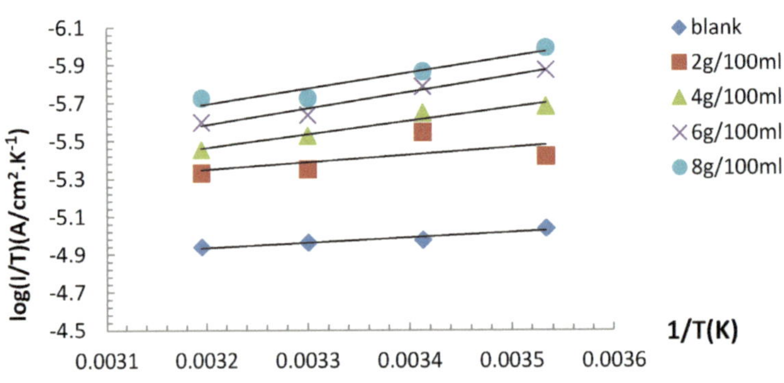

Figure 10.
Variation of log (I_{corr}/T) versus 1/T at various concentrations of TRE.

Inhibitor	C (g/100 mL)	Ea (kJ/mol)	ΔH* (kJ/mol)	ΔS* (J/mol)
0.5 M HCl	—	7.69	5.22	−275.37
Turmeric root extract (TRE)	2.0	10.02	7.55	−275.83
	4.0	15.99	13.52	258.93
	6.0	18.99	16.52	−251.65
	8.0	18.43	15.96	−255.51

Table 4.
Values of activation parameters ΔS and ΔH* for iron 0.5MHCl in the presence and absence of various inhibition concentrations.*

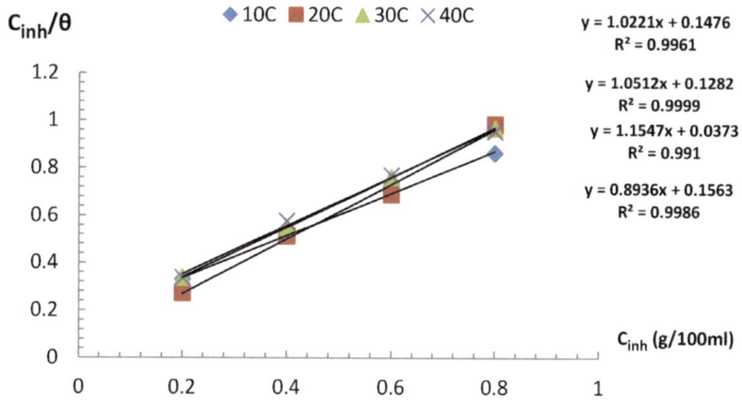

Figure 11.
Plots of Langmuir adsorption isotherm of TRE on iron surface at different temperatures.

pyrolysis process is endothermic. The decreasing entropy of the reaction at the formation of the complex indicates that the adsorbent inhibitor complex has a coherent structure [26–28].

The study of adsorption behavior of active compounds on the iron surface shows the following data:

3.3.3 Adsorption isotherm

Study the effect of turmeric root extract concentration on the inhibition efficacy by acting C\θ in terms of C. Gibbs standard free energy (ΔG_{ads}) is calculated as a single molecule of water, which replaces a molecule. One of the inhibitor molecules has a ratio of 1/55.5, from relationships (13) and (14), according to the isotope of the immersion in adsorption. Langmuir adsorption isotherm is represented by Eq. (13) [29]:

$$\text{Log K} = \log\ (1\backslash 55.5) - \Delta G^{0}{}_{ads}\backslash RT \tag{13}$$

Rearrangement gives Eq. (14):

$$C\backslash\theta = 1/K_{ads} + C \tag{14}$$

where θ is the degree of surface cover with the inhibitor, K_{ads} is the adsorption equilibrium constant, and C is the concentration of inhibitor used in the corrosive medium (**Figure 11**).

T (K)	1/K	ΔG_{ads} (kJ mol^{-1})	ΔSa (J mol^{-1})	R^2	ΔHa (kJ mol^{-1})
283	0.16	−13.8169	48.75	0.9986	−20
293	0.04	−17.7954	60.67	0.9910	
303	0.13	−15.2926	50.40	0.9999	
313	0.15	−15.4306	49.24	0.9961	

Table 5.
Calculated parameters of Langmuir adsorption isotherm.

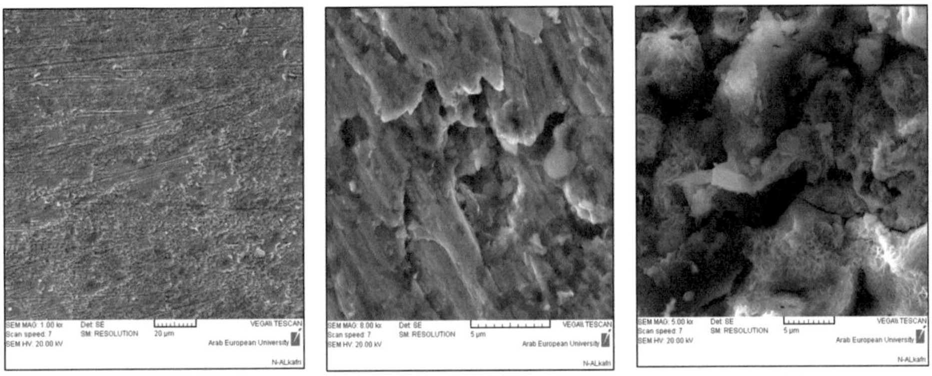

(a) (b) (c)

Figure 12.
SEM of polished iron (a) before immersion (b) after 1 hour of immersion in HCl 1 g/100 mL of TRE and (c) treated iron in the presence of 1 g/100 mL extract.

The value of ΔG_a is less than −20 kJ mol^{-1} which is an indication that physical adsorption is dominant [30] (**Table 5**).

3.4 SEM–EDX analysis

Surface morphology of iron was studied by scanning electron microscopy after 1 h immersion in HCl with and without addition of the inhibitor. **Figure 12a** represents the micrograph obtained by polished steel before exposing to the corrosive medium, while **Figure 12b** showed strongly damaged steel surface due to the corrosion effect after immersion in HCl solution. SEM images of steel surface after 1 hour immersion in HCl with 1 g/100 mL TRE are shown in **Figure 12c**. It can be seen from **Figure 12a** that the iron sample before immersion seems smooth and shows some abrading scratches on the surface. Inspection of **Figure 12b** reveals that the iron surface after immersion in uninhibited HCl shows an aggressive attack of the corroding medium on the iron surface. In contrast, in the presence of 1 g/100 mL TRE (**Figure 12c**), the iron surface was corroded only negligibly. In addition, there was an adsorbed film on the iron surface that was not observed in **Figure 12b**. These results confirmed enhancement of surface coverage of steel surface that led to

Element	C	O	Cl
Without inhibitor	0.20	—	—
After adding HCl	0.20	29.86	1.76
With inhibitor	26.84	7.14	0.1

Table 6.
The percentages of the studied elements in the presence of TRE.

DOI: http://dx.doi.org/10.5772/intechopen.92648

the decrease in contact between the iron and the aggressive medium. Thus, a good adsorptive protection layer that was formed by the inhibitor can efficiently inhibit corrosion of steel.

The following table shows the percentages of the studied elements in the presence and absence of TRE.

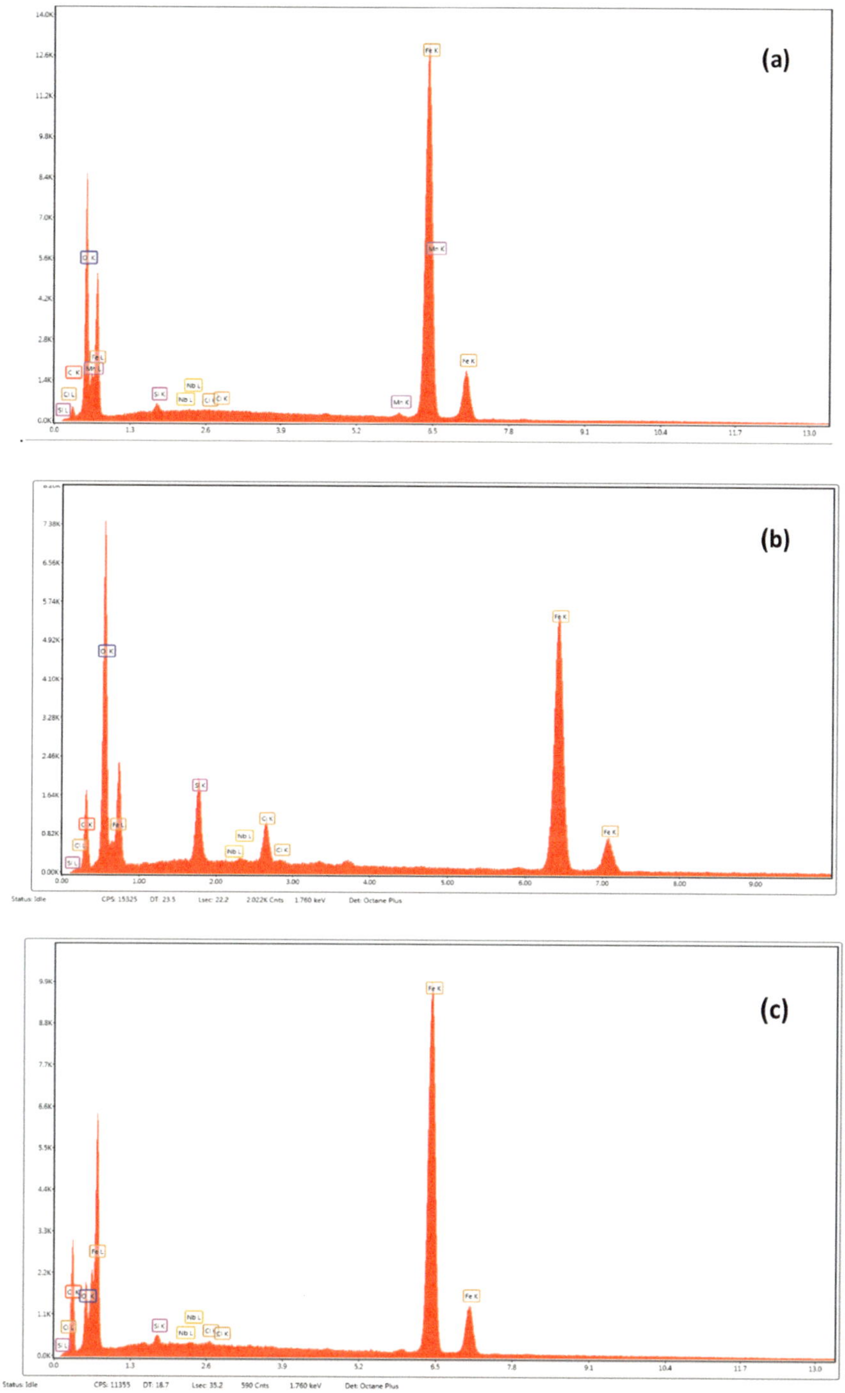

Figure 13.
EDX of (a) polished iron, (b) after 1 hour of immersion in HCl, and (c) treated iron in the presence of 1 g/100 mL extract.

Increasing the percentage of carbon on the surface of the iron treated with the inhibitor compared to the surface of the polished iron is presented in **Table 6** and **Figure 13**, due to the spread of the effective groups in the inhibitor and the adsorption of its electrons on the surface of the iron. Oxygen content is reduced at the surface of the iron treated with an inhibitor, compared to the surface of iron exposed to the corrosive medium, due to its interaction with iron to form a complex inhibitor film. Chlorine content is reduced in the presence of the extract compared to the iron surface exposed to corrosive medium due to the reaction of chlorine with iron forming chloride dissolved iron.

4. Conclusion

The inhibition efficiency (%IE) of TRE increases with the increase of extract concentration. EIS results showed that the double-layer capacitance (Cdl) decreases and charge transfer resistance (Rct) increases with the time of immersion in the extract. The inhibitor showed maximum inhibition efficiency (%IE), 88.9% at 8 g/100 mL concentration. The inhibition efficiency (%IE) of TRE decreased with temperature, which leads to a decrease in activation energy (E_a) of the corrosion process.

The activation energy value of $E_a = -20$ kJ mol^{-1} indicates that the adsorption process is spontaneous and the is physical adsorption. Langmuir adsorption isotherm and SEM studies showed that TRE inhibitions occur through adsorption mechanism. The results of SEM and EDX have been shown to form a protective film on the iron surface.

Acknowledgment

I thank everyone who helped in the face of this work, from laboratory professors to the Atomic Energy Commission.

Author details

Khuloud Almzarzie*, Ayman Almassri, Ahmad Falah and Hassan Kellawi
Department of Chemistry, Damascus University, Syria

*Address all correspondence to: kholod.m990@gmail.com

References

[1] Ahmed A, Kadhum AH, Abdulhadi K. Inhibition of mild steel corrosion in sulfuric acid solution by new Schiff Base. In: Process Engineering. Vol. 7. Malaysia: University Kebangsaan; 2014. pp. 787-804

[2] Ferreira ES, Giacomelli C, Giconelli FC, Spinelli A. Evaluation of the inhibitor effect of L-ascorbic acid on the corrosion of mild steel. Materials Chemistry and Physics. 2004;**83**:129-134

[3] Bentiss F, Gassama F, Barbry D, Gengembre L, Vezin H, Lagrenée M, et al. Corrosion inhibition of mild steel in acidic media using newly synthesized heterocyclic organic molecules: Correlation between inhibition efficiency and chemical structure. Applied Surface Science. 2006;**252**

[4] Lagrenee M, Mernari B, Chaibi N, Traisnel M, Vezin H, Bentiss F. Corrosion inhibition of mild steel in acidic media using newly synthesized heterocyclic organic molecules: Correlation between inhibition efficiency and chemical structure. Corrosion Science. 2001;**43**

[5] Ayers RC, Hackerman N Jr. Corrosion inhibition in HCl using methyl pyridines. Journal of the Electrochemical Society. 1963;**110**(6): 507-513

[6] Moretti G, Guidi F, Grion G. Tryptamine as a green iron corrosion inhibitor in 0.5M deaerated sulphuric acid. Corrosion Science. 2004;**46**(2): 387-403

[7] Quraishi MA, Rawat J, Ajamal M. Dithiobiurets: A novel class of acid corrosion inhibitors for mild steel. Journal of Applied Electrochemistry. 2000;**30**(6):745-751

[8] Khalil N. Quantum chemical approach of corrosion inhibition. Electrochimica Acta. 2003;**48**: 2635-2640

[9] Kairi N. The Effect of Temperature on the Corrosion Inhibition of Mild Steel in 1 M HCl Solution by *Curcuma longa* Extract. Malaysia: Universiti Sains; 2013. pp. 7138-7155

[10] Jinendra G. Corrosion Inhibition by Turmeric Extract. India: RGPV University; 2012. pp. 10455-10458

[11] Aprael S et al. Corrosion Inhibition of Mild Steel by Curcuma Extract in Petroleum Refinery Wastewater. Iraq: University of Baghdad; 2014

[12] Saleh K et al. Inhibition and Adsorption Actions of Nano Curcumin for Corrosion of Carbon Steel Alloy in 3.5% NaCl Solution. Baghdad, Iraq: University of Baghdad, Collage of Science; 2017. pp. 515-529

[13] Johnsirani V et al. Curcumin dye as corrosion inhibitor for carbon steel in sea water. Chemical Science Transactions. 2013

[14] Benali O, Larabi L, Traisnel M, Gengembra L, Harek Y. Study on the inhibition of mild steel corrosion by quaternary ammonium compound in H2SO4. Applied Surface Science. 2007; **253**:6130

[15] Abdel-Gaber AM, Hijazia KM, Younesa GO, Nsouli B. Comparative study of the inhibitive action between the bitter orange leaf extract and its chemical constituent linalool on the mild steel corrosion in HCl solution. Química Nova. 2017;**230**

[16] Mohamed HA, Farag AA, Badran BM. Corrosion Inhibition of Mild Steel Using Emulsified Thiazole

Adduct in Different Binder Systems. Vol. 10. Cairo, Egypt: Department of Polymers and Pigments, National Research Center; 2008. pp. 67-77

[17] Abdel Hamid Z. Clarification of the corrosion inhibition of mild steel in hydrochloric acid solutions via Cetyltrimethyl ammonium bromide inhibitor. Egyptian Journal of Chemistry. 2014;**57**(5):353-371

[18] Shukla SK, Quraishi MA. The effects of pharmaceutically active compound doxycycline on the corrosion of mild steel in hydrochloric acid solution (HCl). Corrosion Science. 2010;**52**:314

[19] Yurt A, Balaban A, Kandemir SU, Bereket G, Erk B. Adsorption and corrosion inhibition characteristics of some nicotinamide derivatives on mild steel in hydrochloric acid solution (HCl). Materials Chemistry and Physics. 2004;**85**

[20] Hui Cang W. Study of Stevia rebaudiana Leaves as Green Corrosion Inhibitor for Mild Steel in Sulphuric Acid by Electrochemical Techniques. College of chemical engineering and biological, Yancheng institute of technology; 2012. pp. 3726-3736

[21] Abd EI-Rehim S, Ibrahim MAM, Khaled KF. 4-Aminoantipyrine as an inhibitor of mild steel corrosion in HCl solution. Journal of Applied Electrochemistry. 1999;**29**

[22] Ch W-M, Post F, Joshua L, Sazzadur R, Anjali T. Sub-Surface Corrosion Research on Rock Bolt System, Perforated SS Sheets and Steel Sets for the Yucca Mountain Repository. Vol. 302004

[23] Bentiss F, Bouanis M, Mernari B, Traisnel M, Vezin H, Lagrenee M. Applied Surface Science. 2007;**253**

[24] Noor EA, Al-Moubaraki AH. Thermodynamic study of metal corrosion and inhibitor adsorption processes in mild steel/1-methyl-4[4′(−X)-styryl] pyridinium iodides/hydrochloric acid systems. Materials Chemistry and Physics. 2008;**110**

[25] Sayyah SM, Abd El-Rehim MM, Mohamed SM. Corrosion inhibition of Aluminium with a series of aniline monomeric surfactant and their analogues polymers in 0.5 M HCl solution. Egyptian Journal of Chemistry. 2012;**55**(6):583-602

[26] Maqsood AM, Ali H, Firdosa N, Shaeel A, Zaheer K. Anti-corrosion ability of surfactants. International Journal of Electrochemical Science. 2011;**6**:1927-1948

[27] Abd El-Rahim SS, Refaey SAM, Taha F, Saleh MB, Ahmed RAJ. Novel cationic surfactants from fatty acids and their corrosion inhibition efficiency for carbon steel pipelines in 1 M HCl. Journal of Applied Electrochemistry. 2001;**31**

[28] Abd El-Rehim SS, Hassan HH, Amin MA. Corrosion inhibition of aluminum in hydrochloric acid solution (HCl) using potassium iodate inhibitor. Materials Chemistry and Physics. 2001;**70**

[29] Bilgic S, Caliskan N. Inhibition of steel corrosion in hydrochloric acid solution (HCl) by chamomile extract. Applied Surface Science. 1999;**152**

[30] Tsuru T, Haruyama S, Gijutsu B. Corrosion monitor based on impedance method; construction and its application to homogeneous corrosion. Japan Society of Corrosion Engineering. 1978;**27**

Chapter 4

Designing and Synthesis of (Cd^{2+}, Li^{+}), Cr^{3+}, Bi^{3+} Doped $CePO_4$ Materials Optical, Electrochemical, Ionic Conductivity Analysis

Salah Kouass, Amor Fadhalaoui, Hassouna Dhaouadi and Fathi Touati

Abstract

Most of the work has been done on the optical properties of the rare earth doped $CePO_4$, so there are few studies on the effect of metal ion doping on $CePO_4$. The doping improves the properties of the compounds and can lead to new properties. **It is the first time, that multi- ionic doping process is used in the $CePO_4$ matrix, in order to improve the ionic conductivity and the electrochemical stability.** The low percentage of (Cd^{2+}, Li^{+}), Cr^{3+}, Bi^{3+} dopant affect the structure showing a weak decrease in the lattice parameters compared to the $CePO_4$. Impedance spectroscopy analysis was used to analyze the electrical behavior of samples as a function of frequency at different temperatures. The total electrical conductivity plots obtained from impedance spectra shows an increase of the total conductivity as Li, Cr-content increases. The determined energy gap values decrease with increasingly Li^{+}, Cr^{3+} and Bi^{3+} doping content. Electrochemical tests showed an improved capacity when increasing the Li^{+}, Cr^{3+} and Bi^{3+} content and a stable cycling performance.

Keywords: phosphate materials, doping, optical properties, impedance spectroscopy, electrochemical properties

1. Introduction

Nanoscience and nanotechnology is a rapid-developing field which has demanded the technologist to innovate applicable nanomaterials with manipulated shape and size to explore their principal chemical and physical characteristics [1]. In recent years, rare earth phosphates have attracted many researchers because of their technological applications [2, 3]. Cerium orthophosphate **nanomaterials** have important properties: high thermal stability [4], **very low solubility in water, their use in the production of moisture sensors for luminescent materials, a poison for automotive catalysts and a novel oxygen sensing material on the basis of its redox responsive reversible luminescence** [5–7].

Most of the work has been done on the optical properties of the rare earth doped $CePO_4$, so there are few studies on the effect of metal ion doping on $CePO_4$. Additionally, $CePO_4$ materials have been used in hydrogen fuel cells [8]. To better understand the mechanism of conduction, information on the behavior and ionic conductivities of charge carriers located in phosphates, electrical studies have been carried out.

Generally, the doping process improves the properties of the compounds and can lead to new properties [9, 10]. Trivalent elements have been known as doping elements, improving the physico-chemical properties of cerium phosphate-based materials [11]. In order to improve the electrical and optical properties, the cerium phosphate was partially substituted by divalent transition metal ions. The doping with Ca and Sr. has improved the electrical conductivity of (La, Ce) PO_4 [12, 13]. The high conductivity of the Sr-doped $CePO_4$ under wet oxidizing conditions due to electronic and ionic conduction is shown by Moral et al. [12]. Norby et al. studied the effect of the substitution of lanthanum by calcium and strontium on the conductivity, described by the dependence on humidity and the effect of H/D isotopic exchange [13].

The substitution effect depends on the nature of the doping elements. Chromium shows the stability of the valence state (+ III) in conductive p-type SOFC interconnection materials [14]. Numerous reports show that substitution with Cr^{3+} ions introduces interesting properties in ferrites [15, 16]. Cr-doping $CePO_4$ is expected to improve its optical and electrical properties.

Bismuth-based materials have been studied because of their excellent photocatalytic activities in the reduction of NO [17], the generation of O_2 [18, 19] and the decomposition of organic compounds [20, 21]. **It was founded that Y_2SiO_5:Bi^{3+} gives rise to three emission bands centering at: 355, 408, and 504 nm upon UV excitation possibly from three types of bismuth emission centers in the compound, respectively** [22]. **The broad absorption band of Bi^{3+} improves the emission process which could be varied from the UV to the NIR, depending on its final valence in the compounds** [23]. The Bi^{3+} ions combined with rare earth ions such as cerium, Ce^{3+}, can improve the optical properties of $CePO_4$ nanomaterials. The study of the effect of doping with Bi^{3+} ions on the structural and electrical properties of $CePO_4$ is virgin. This leads to new optical and electrical properties for application in electronic devices.

Divalent cations were doped in monophosphates, giving variations in the electrical properties of these doped materials. The aim is to study the combined effect of monovalent Li^+ and divalent Cd^{2+} ions on structural, electrical and optical properties. Indeed, the electrical and electrochemical properties of cadmium allow it to be used in mobile phone batteries [24, 25]. Also lithium Li^+ ions associated with the divalent Fe^{2+}, Mn^{2+} and Co^{2+} ions favor the increase of the capacity, the lifetime, **diffusion process** and the electrochemical stability of a phosphate-based electrode [26–28]. **The adjustment of the size, shape, density, optical, electrical and dielectric properties of nanoparticles could help tune their broad spectral resonance wavelength [29]. Microemulsion approach associated to the hydrothermal conditions could be used to fabricate single crystalline $CePO_4$ nanowires with controlled aspect ratios [30]. Hydrothermal process has emerged as a powerful tool due to some significant advantages such as cost-effective, controllable particle size, low-temperature and less-complicated techniques [31].**

2. Characterizations

Cerium orthophosphate has two crystalline phases [32, 33]. At low temperature this material crystallizes in the hexagonal system. At high temperature cerium orthophosphate crystallizes in the monoclinic system. The hexagonal structure is characterized by the existence of large tunnels parallel to the c-axis in which the water present in the compound appears to be localized. The $CePO_4$ produced in aqueous solution at room temperature crystallizes in a hexagonal form [34, 35]. After heat treatment at 650°C, the hexagonal phase ($CePO_4$) started converting into a monoclinic structure.

The ions (Cd^{2+}, Li^{+}), Cr^{3+}, Bi^{3+} doped $CePO_4$ materials were characterized by X-ray diffraction (XRD). All samples are single phase having a hexagonal structure similar to $CePO_4$. **The** 2θ values of doped materials shift slightly higher angles with increasing Cr, Bi, Cd and Li content, confirming the complete dissolution of dopants (**Figure 1**). The same behavior was observed when Fe^{3+} ion substitutes La^{3+} ion in $LaPO_4$ [36]. The average crystallite size of all samples decreases with increasing the amount of doping. The main reason for the decrease of the grain size may be due to the fact that doping introduced defects and the defects prevent grain to grow [37].

Many parameters affecting the morphological characteristics of the hexagonal cerium phosphate nanocrystals such as the cerium concentration, the treatment temperature, the reaction time, the nature of the surfactant, the pH value of the solution and the synthesis method. The materials take on a similar shape to the nanorod morphology with the size depending on the dopant-content.

3. Optical properties

The band gap energy of the as-prepared samples was calculated using the Kubelka-Munk plot. The Kubelka-Munk function for diffuse reflectance [38] is

$$f(R) = \frac{1 - R^2}{2.R} \quad (1)$$

where R is the reflectance. The optical band gap, E_g, can be determined using the Tauc relation:

$$[F(R).h\nu] = A[h\nu - E_g]^n \quad (2)$$

where A is an energy-independent constant, E_g is the optical band gap and n can take values of 0.5, 1.5, 2 and 3 depending on the mode of transition [39]. The band gap energies can be estimated by extrapolating the linear portions to the hν axis and from the corresponding intercept of the tangents to the plots of $[F(R)*h\nu]^2$ vs. hν.

The determined energy gap values decrease with increasing Cr, Bi, Cd and Li-doping content in $Cr_xCe_{1-x}PO_4$ (x = 0.00, 0.08, 0.10 and 0.20), $Bi_xCe_{1-x}PO_4$ (x = 0.00, 0.02 and 0.08), $Ce_{0.9}Cd_{0.15-x}Li_{2x}PO_4$ (x = 0 and 0.02) nanorods,

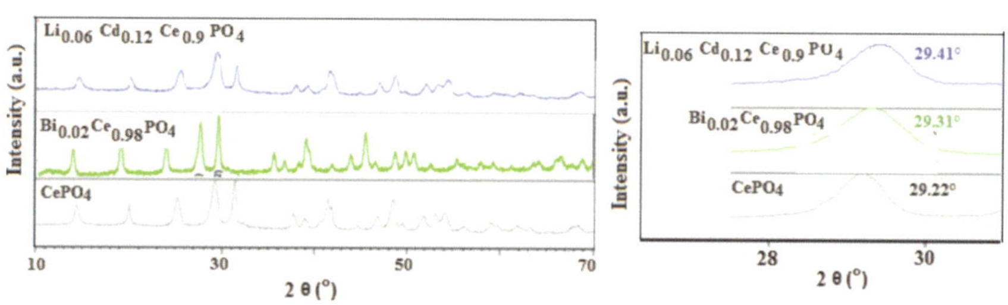

Figure 1.
X-ray diffraction pattern of $CePO_4$, $Bi_{0.02}Ce_{0.98}PO_4$ and $Li_{0.06}Cd_{0.12}Ce_{0.90}PO_4$.

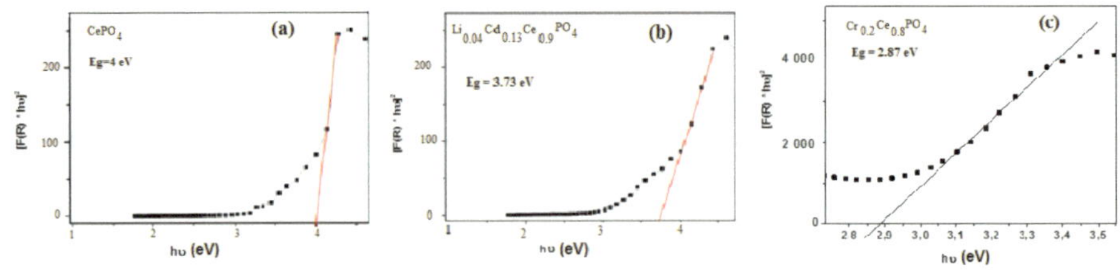

Figure 2.
$[f(R) \times h\nu]2$ versus the $h\nu$ (eV) plots of: (a) $CePO_4$; (b): $Ce_{0.9}Cd_{0.13}Li_{0.04}PO_4$; and (c) $Cr_{0.20}Ce_{0.80}PO_4$.

$Cr_xCe_{1-x}PO_4$ Eg (eV) [40]	$Bi_xCe_{1-x}PO_4$ Eg (eV) [41]	$Ce_{0.9}Cd_{0.15-x}Li_{2x}PO_4$ Eg (eV) [42]
$CePO_4$ Eg = 4.14	$CePO_4$ Eg = 4.00	$CePO_4$ Eg = 4.00
$Cr_{0.08}Ce_{0.92}PO_4$ Eg = 4.10	$Bi_{0.02}Ce_{0.98}PO_4$ Eg = 3.96	$Ce_{0.9}Cd_{0.15}PO_4$ Eg = 3.95
$Cr_{0.10}Ce_{0.90}PO_4$ Eg = 3.09	$Bi_{0.08}Ce_{0.92}PO_4$ Eg = 3.84	$Ce_{0.9}Cd_{0.13}Li_{0.04}PO_4$ Eg = 3.73
$Cr_{0.20}Ce_{0.80}PO_4$ Eg = 2.87		

Table 1.
Gap energy values of $Cr_xCe_{1-x}PO_4$, $Bi_xCe_{1-x}PO_4$ and $Ce_{0.9}Cd_{0.15-x}Li_{2x}PO_4$ nanomaterials.

respectively, showing a red-shift trend when the doping- substitution percentage increases (**Figure 2**). **Table 1** summarizes the gap energy values of nanomaterials.

The size, morphology and substitution of crystallites affect the energy of the band gap. The substitution of Ce^{3+} by a transition metal could induce the formation of several structural defects, creating different energy levels below the conduction band. The same behavior has been observed in Cr-doped $Ni_3(PO_4)_2$ where the band gap decreases when Cr^{3+} replaces Ni^{2+} [43].

4. Electrical conductivity

The dc-conductivity (σ_{dc}) of $Bi_xCe_{1-x}PO_4$ could be calculated using the Formula's

$$\sigma_{dc} = \frac{t}{A} * \frac{1}{R} \tag{3}$$

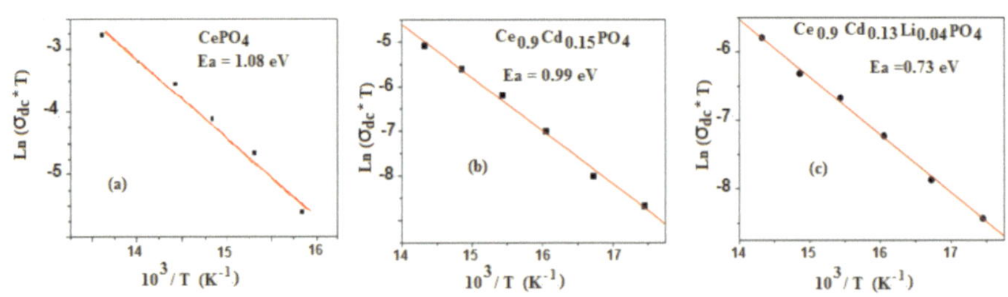

Figure 3.
Arrhenius plot of the electrical conductivity of $CePO_4$, $Ce_{0.9}Cd_{0.15}PO_4$ and $Ce_{0.9}Cd_{013}Li_{0.04}PO_4$.

$Cr_xCe_{1-x}PO_4$ Ea (eV) [40]	$Bi_xCe_{1-x}PO_4$ Ea (eV) [41]	$Ce_{0.9}Cd_{0.15-x}Li_{2x}PO_4$ Ea (eV) [42]
$CePO_4$ Ea = 1.08	$CePO_4$ Ea = 0.84	$CePO_4$ Ea = 1.08
$Cr_{0.08}Ce_{0.92}PO_4$ Ea = 0.90	$Bi_{0.02}Ce_{0.98}PO_4$ Ea = 0.87	$Ce_{0.9}Cd_{0.15}PO_4$ Ea = 0.99
$Cr_{0.10}Ce_{0.90}PO_4$ Ea = 0.84	$Bi_{0.08}Ce_{0.92}PO_4$ Ea = 1.09	$Ce_{0.9}Cd_{0.13}Li_{0.04}PO_4$ Ea = 0.72
$Cr_{0.20}Ce_{0.80}PO_4$ Ea = 0.80		

Table 2.
Activation energy of $Cr_xCe_{1-x}PO_4$, $Bi_xCe_{1-x}PO_4$ and $Ce_{0.9}Cd_{0.15-x}Li_{2x}PO_4$.

(A = area of the sample surface and t = sample thickness). The temperature dependence of dc-conductivity could be plotted based on the Arrhenius law with the following expression:

$$\sigma_{dc} = \frac{A_0}{T} e^{-\frac{Edc}{K.T}} \tag{4}$$

where A_0 is the pre-exponential factor, Ea the activation energy and K the Boltzmann constant.

The activation energy of the undoped $CePO_4$ nanorods (Ea = 1.08 eV) is comparable to that obtained for $CePO_4$ nanosheets (Ea = 1.06 eV) [44]. It seems that the change of the morphology and the synthesis route used weakly affect the activation energy of the cerium phosphates. The activation energy deduced from Log (σT) as a function of 10^3/T (**Figure 3**) are summarized in **Table 2**.

The effect of Cr^{3+}, (Cd^{2+}, Li^+) substitutions decreases the activation energies with the increase in Cr, (Cd, Li)-concentration (**Table 2**). Consequently, the dc-conductivity of the as-prepared samples increases with temperature and with doping concentration. Lattice defects and distortions in the phosphate structure produced by the substitution allow the increase of the DC conductivity. The enhancement of activation energy could be related to the mobility of oxygen ions (O_2^-). This phenomenon has been observed by Nandini et al. [45]. They show that with an appropriate ratio of magnesium and strontium, the ionic conductivity increases as compared to that exhibited by ceria singly doped with Mg.

The difference in the electrical transport process between the Cr, Cd, Li doped $CePO_4$ and the Bi-doped $CePO_4$ results from the difference in atomic weight of Bi and Cr, Cd, Li. The atomic weight affects the mobility of the ions and therefore the Bi^{3+} ions remain close to their initial positions.

5. Electrochemical measurements

In order to explore the potential application of nonmaterials as cathode materials, their electrochemical performance with respect to Li insertion/extraction was investigated. Cyclic voltammograms (CVs) for $CePO_4$, $Ce_{0.9}Cd_{0.15}PO_4$ and $Bi_{0.02}Ce_{0.98}PO_4$ nanorods (examples) at 20 mV/s are shown in **Figure 1**. For all the as-prepared compounds, the cyclic voltammograms are well superposed indicating the relative structural stability under these conditions. The same shape of the CV curves slightly is observed for Nanoplate-like CuO in the presence of $LiClO_4$ in propylene carbonate [46].

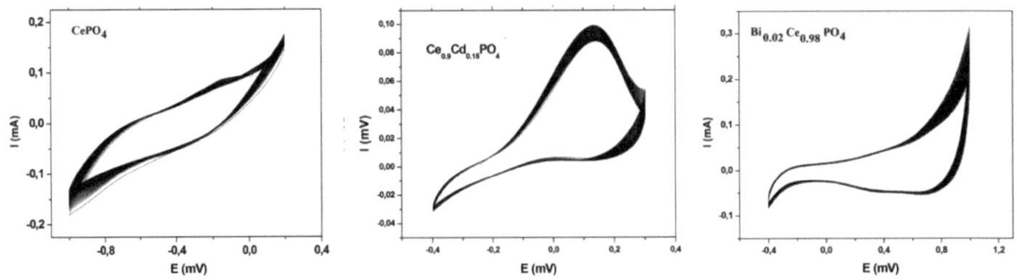

$CePO_4$, $Cd_{0.15}$ $Ce_{0.90}PO_4$ and $Bi_{0.02}Ce_{0.98}PO_4$ based electrode cyclic voltammogramm.

These voltammograms indicate the intercalation/de-intercalation process of Li^+ ions. During the electrochemical redox processes, the intercalation/de-intercalation process of Li^+ ions can be represented by the following reaction:

$$M_xCe_{1-x}PO_4 + ye^- + yLi^+ \leftrightarrow Liy\ M_xCe_{1-x}PO_4$$

$$\text{Intercalation of Li} \rightarrow$$

$$\leftarrow \text{De} - \text{intercalation of Li}$$

The lithium ion diffusion coefficients can be calculated from the Randles-Sevcik law [47]:

$$ip = (2.69 \times 10^5)\ n^{(3/2)}.\ A\ C\ D_{Li}^{1/2}\ v^{1/2} \tag{5}$$

where ip is the peak current (A), n is the number of electrons exchanged, A is the apparent surface area of the electrode (cm^2), D_{li} and C are the diffusion coefficient (cm^2/s) and the analyte concentration (in moles/cm^3) respectively, and V is the potential scan rate (V/s). The lithium ion diffusion coefficients deduced are 2.5×10^{-9}, 0.7×10^{-9}, 4.6×10^{-9} cm^2s^{-1} for $CePO_4$, $Ce_{0.9}Cd_{0.15}PO_4$ and $Ce_{0.9}Cd_{0.13}Li_{0.04}PO_4$, respectively. The structure, surface area, grain size and morphology affect the calculated lithium diffusion coefficient DLi of the electrode materials. For example, Bi doping with the appropriate amount improved the electrochemical performance of $LiFePO_4$ cathode material, synthesized by the sol–gel method [48].

For as-prepared $Bi_xCe_{1-x}PO_4$ (x = 0.00, 0.02, 0.08) electrodes, The lithium ion diffusion coefficient (D_{Li}) values could be determined by using Nyquist plot through the relation [49]:

$$D_{Li} = \frac{R^2\ T^2\ V_M^2}{2A^2\ n^4\ F^4\ \sigma^2} \tag{6}$$

Where: F, R and T indicate Faraday constant, gas constant and room temperature, respectively.

(1). D_{Li} can be calculated as the Warburg impedance Z_w is inversely proportional to the square root of the diffusion coefficient as shown in [50]. The calculated lithium diffusion coefficient of the $CePO_4$ and $Bi_{0.02}Ce_{0.98}PO_4$ and $Bi_{0.08}Ce_{0.92}PO_4$ electrodes is 3.3×10^{-16}, 40×10^{-16} and 12.8×10^{-16} $cm^2.s^{-1}$ respectively. The D_{Li} variation values n can be attributed to creating the defect and increasing disorder of the lattice in doped $CePO_4$, drives to the improvement of the electrochemical performance. The structure of H-$CePO_4$-type characterized by infinite tunnels

Specific capacitances C (Fg^{-1}) [41]	Specific capacitances C (Fg^{-1}) [42]
$CePO_4$ C = 58	$CePO_4$ C = 58
$Bi_{0.02}Ce_{0.98}PO_4$ C = 63	$Ce_{0.9}Cd_{0.15}PO_4$ C = 76
$Bi_{0.08}Ce_{0.92}PO_4$ C = 75	$Ce_{0.9}Cd_{0.13}Li_{0.04}PO_4$ C = 120

Table 3.
Specific capacitances of $Bi_xCe_{1-x}PO_4$ and $Ce_{0.9}Cd_{0.15-x}Li_{2x}PO_4$ nanomaterials.

provides fast ionic transport. The Li + ions can move quickly in an appropriate direction [51].

The specific capacitance can be estimated by the following equation [52, 53]:

$$C = \frac{\int Idv}{s.w.\Delta V} \quad (7)$$

where ΔV is the potential window, m is the mass of active material in one electrode, I is the current, and s is the potential scan rate. The variation of the specific capacitance of two prepared simples versus cycle number is given in **Table 3.** We show that the partial substitution of Ce by Cd increase the capacitance. The increasing of the capacitance can be attributed to the partial substitution and the small crystal size which improves the kinetics of electrochemical reactions and the structure which provides fast ionic transport.

The reason for the improvement of the discharge capacity can be explained as follows: with Bi-doping, the grain size of the particles decreases, which leads to the migration of the Li-ion.

The penetration of electrolyte ions and the electrochemical activation of the materials may increase the specific capacitance. A similar phenomenon has been observed by other authors [54, 55].

Doped samples show better performance in terms of discharge capacity than undoped ones. These results could be attributed to the contribution of the nanorod shape and the particle size. Indeed, the reduction of the size allows a faradic reaction providing a short ion diffusion path and electron transport.

6. Conclusion

In summary, we have demonstrated a rapid and convenient hydrothermal method for the preparation of doped and undoped $CePO_4$ nanomaterials. The Cr^{3+}, Bi^{3+}, Cd^{2+} and Li^+ ions substitution affects the optical, electrical and electrochemical properties. The band gap energies of the as-prepared $CePO_4$ nanorods decreased with increasing doping-concentration showing a red-shift trend. Comparative experiments have witnessed that the doped-$CePO_4$ electrode had the most excellent electrochemical properties in comparison with undoped $CePO_4$ nanomaterials. The electrochemical results show that the specific capacity and the electrical conductivity increase with increasing doping content. **The specific capacitance of the hybrid electrode materials presents a good cyclic stability. The improved specific capacitance is due to the surface morphology and the decrease of grain size of the particles. The lowering in the crystal size allows a fast faradaic reaction, giving a short ion diffusion path, which improves the electrochemical properties. This simple synthesis methodology together with the good optical and electronic properties makes this material scientifically; technologically interesting and could find a potential use in nanoelectronics.**

Author details

Salah Kouass[1]*, Amor Fadhalaoui[2], Hassouna Dhaouadi[3] and Fathi Touati[3]

1 Laboratoire Matériaux Utiles, Institut National de Recherche et d'Analyse Physico-chimique (INRAP), Sidi Thabet, Tunis, Tunisie

2 Laboratoire de Physique des Matériaux, Faculté des Sciences de Bizerte, Université de Carthage, Zarzouna, Bizerte, Tunisia

3 Laboratoire Matériaux Traitement et Analyse, Institut National de Recherche et d'Analyse Physico-chimique (INRAP), Sidi Thabet, Tunis, Tunisie

*Address all correspondence to: koissa2000@yahoo.fr

References

[1] Asiya SI, Kaushik P, Gharieb El-S, Abd Elkodous M, Demetriades C, Kralj S, et al. Reliable optoelectronic switchable device implementation by CdS nanowires conjugated bent-core liquid crystal matrix. Organic Electronics. 2020;**82**:10559

[2] Ho LN, Nishiguchi H, Nagaoka K, Takita Y. Synthesis and characterization of a series of mesoporous nanocrystalline lanthanides phosphate. Journal of Porous Materials. 2006;**13**: 237-244

[3] Zhang YJ, Guan HM. The growth of lanthanum phosphate (rhabdophane) nanofibers via the hydrothermal method. Materials Research Bulletin. 2005;**40**:1536-1543

[4] Hikichi Y, Nomura T, Tanimura Y, Sb S. Sintering and properties of monazite-type $CePO_4$. Journal of the American Ceramic Society. 1990;**73**: 3594-3596

[5] Xu L, Guo G, Uy D, O'Neill AE, Weber WH, Rokosz MJ, et al. Cerium phosphate in automotive exhaust catalyst poisoning. Applied Catalysis B: Environmental. 2004;**50**:113-125

[6] Granados ML, Galisteo FC, Lambrou PS, Mariscal R, Sanz J, Sobrados I, et al. Role of P-containing species in phosphated CeO_2 in the deterioration of its oxygen storage and release properties. Journal of Catalysis. 2006;**239**:410-421

[7] Di W, Wang X, Ren X. Nanocrystalline $CePO_4$:Tb as a novel oxygen sensing material on the basis of its redox responsive reversible luminescence. Nanotechnology. 2010; **21**:075709

[8] Kitamura N, Amezawa K, Tomii Y, Hanada T, Yamamoto N, Omata T, et al. Electrical conduction properties of Sr-doped $LaPO_4$ and $CePO_4$ under oxidizing and reducing conditions. Journal of the Electrochemical Society. 2005;**152**:A 658-A 663

[9] Feng X, Cheng Y, Ye C, Ye J, Peng J, Hu J. Synthesis and Ag-content-depended electrochemical properties of Ag/ZnO heterostructured nanomaterials. Materials Letters. 2012; **79**:205-208

[10] Hu J, Yu Y, Guo H, Chen Z, Li A, Feng X, et al. Sol–gel hydrothermal synthesis and enhanced biosensing properties of nanoplated lanthanum-substituted bismuth titanate microspheres. Journal of Materials Chemistry. 2011;**21**:5352-5359

[11] Yang M, You H, Liu K, Zheng Y, Guo N, Zhang H. Low-temperature coprecipitation synthesis and luminescent properties of $LaPO_4:Ln^{3+}$ (Ln^{3+} = Ce^{3+}, Tb^{3+}) nanowires and $LaPO_4:Ce^{3+},Tb^{3+}/LaPO_4$ core/shell nanowires. Inorganic Chemistry. 2010; **49**:4996-5002

[12] del Moral EG, Fagg DP, Chinarro E, Abrantes JCC, Jurado JR, Mather GC. Impedance analysis of Sr-substituted $CePO_4$ with mixed protonic and p-type electronic conduction. Ceramics International. 2009;**35**:1481-1486

[13] Norby T, Christiansen N. Proton conduction in Ca- and Sr-substituted $LaPO_4$. Solid State Ionics. 1995;**77**:240-243

[14] Ding X, Liu Y, Gao L, Guo L. Effects of cation substitution on thermal expansion and electrical properties of lanthanum chromites. Journal of Alloys and Compounds. 2006;**425**:318-322

[15] Yunus SM, Yamauchi H, Zakaria H, Igawa N, Hoshikawa A, Ishii Y. Neutron diffraction studies of the magnetic ordering in the spinel oxide system

$Mg_xCo_{1-x}Cr_xFe_{2-x}O_4$. Journal of Alloys and Compounds. 2008;**455**:98-105

[16] Wahba AM, Mohamed MB. Structural, magnetic, and dielectric properties of nanocrystalline Cr-substituted $Co_{0.8}Ni_{0.2}Fe_2O_4$ ferrite. Ceramics International. 2014;**40**: 6127-6135

[17] Li G, Zhang D, Yu JC, Leung MKH. An efficient bismuth tungstate visible-light-driven photocatalyst for breaking down nitric oxide. Environmental Science & Technology. 2010;**44**: 4276-4281

[18] Bessekhouad Y, Mohammedi M, Trari M. Hydrogen photoproduction from hydrogen sulfide on Bi_2S_3 catalyst. Solar Energy Materials & Solar Cells. 2002;**73**:339-350

[19] ShimodairaY KH, Kobayashi H, Kudo A. Photophysical properties and photocatalytic activities of bismuth molybdates under visible light irradiation. The Journal of Physical Chemistry. B. 2006;**110**:17790-17797

[20] Xiong J, Cheng G, Lu Z, Tang J, Yu X, Chen R. BiOCOOH hierarchical nanostructures: Shape-controlled solvothermal synthesis and photocatalytic degradation performances. CrystEngComm. 2011;**13**: 2381-2390

[21] Tang J, Zou Z, Ye J. Efficient photocatalytic decomposition of organic contaminants over $CaBi_2O_4$ under visible-light irradiation. Angewandte Chemie, International Edition. 2004;**43**: 4463-4466

[22] Kang F, Zhang Y, Peng M. Controlling the energy transfer via multi luminescent centers to achieve white light/tunable emissions in a single-phased X2-Type $Y_2SiO_5:Eu^{3+}$, Bi^{3+} phosphor for ultraviolet converted LEDs. Inorganic Chemistry. DOI: 10.1021/ic502439k

[23] Kang F, Peng M, Zhang Q, Qiu J. Abnormal anti-quenching and controllable multi-transitions of Bi^{3+} luminescence by temperature in a yellow-emitting $LuVO_4:Bi^{3+}$ phosphor for UV-converted white LEDs. Chemistry - A European Journal. 2014; **20**:11522-11530

[24] Arabzadeh A, Salimi A. One dimensional CdS nanowire@TiO_2 nanoparticles core-shell as high performance photocatalyst for fast degradation of dye pollutants under visible and sunlight irradiation. Journal of Colloid and Interface Science. 2016; **479**:43

[25] Lee SM, Yeon DH, Chon SS, Cho YS. Effect of double substitutions of Cd and Cu on optical band gap and electrical properties of non-colloidal PbS thin films. Journal of Alloys and Compounds. 2016;**685**:129

[26] Xiao Y, Chun F, Zhang J, Han I. Electrical structures, magnetic polaron and lithium ion dynamics in three transition metal doped $LiFe_{1-x}M_xPO_4$ (M = Mn, Co and La) cathode material for Li ion batteries from density functional theory study. Solid State Ionics. 2016;**294**:73-81

[27] Yuan H, Wang X, Wu Q, Shu H, Yang X. Effects of Ni and Mn doping on physicochemical and electrochemical performances of $LiFePO_4$/C. Journal of Alloys and Compounds. 2016;**675**:187-194

[28] Zhou H, Upreti S, Chernova NA, Hautier G, Ceder G, Whittingham MS. Iron and manganese pyrophosphates as cathodes for lithium-ion batteries. Chemistry of Materials. 2011;**23**:293

[29] Thirugnanasambandan T, Pal K, Sidhu A, Elkodous MA, Prasath H, Kulasekarapandian K, et al. Aggrandize efficiency of ultra-thin silicon solar cell via topical clustering of silver nanoparticles. Nano-Structures & Nano-Objects. 2018;**16**:224-233

[30] Cao M, Hu C, Wu Q, Guo C, Qi Y, Wang E. Controlled synthesis of $LaPO_4$ and $CePO_4$ nanorods/nanowires. Nanotechnology. 2005;**16**:282-286

[31] Pala K, Maitia UN, Majumdera TP, Debnath SC. A facile strategy for the fabrication of uniform CdS nanowires with high yield and its controlled morphological growth with the assistance of PEG in hydrothermal route. Applied Surface Science. 2011; **258**:163-168

[32] Bao J, Yu R, Zhang J, Yang X, Wang D, Deng J, et al. Low-temperature hydrothermal synthesis and structure control of nano-sized $CePO_4$. CrystEngComm. 2009;**11**:1630

[33] Ma L, Chen W-X, Zheng Y-F, Xu Z-D. Hydrothermal growth and morphology evolution of $CePO_4$ aggregates by a complexing method. Materials Research Bulletin. 2008;**43**: 2840

[34] Yan RX, Sun XM, Wang X, Peng Q, Li YD. Crystal structures, anisotropic growth, and optical properties: Controlled synthesis of lanthanide orthophosphate one-dimensional nanomaterials. Chemistry - A European Journal. 2005;**11**:2183

[35] Zollfrank C, Scheel H, Brungs S, Greil PJ. Europium(III) orthophosphates: Synthesis, characterization, and optical properties. Crystal Growth & Design. 2008;**8**:766

[36] Guo D, Hu C, Xi Y. Synthesis and magnetic property of Fe doped $LaPO_4$ nanorods. Applied Surface Science. 2013;**268**:458-463

[37] Wu L, Wang Z, Li X, et al. Electrochemical performance of Ti^{4+}-doped $LiFePO_4$ synthesized by co-precipitation and post-sintering method. Transactions of the Nonferrous Metals Society of China. 2010;**20**: 814-818

[38] Miyake Y, Tada H. Photocatalytic degradation of methylene blue with metal doped mesoporous titania under irradiation of white light. Journal of Chemical Engineering of Japan. 2004; **37**:630-635

[39] Tauc J, Menth A. States in the gap. Journal of Non-Crystalline Solids. 1972; **8–10**:569-585

[40] Fadhalaouia A, Dhaouadib H, Marouania H, Koukic A, Madanid A, Rzaiguia M. Cr-substitution effect on structural, optical and electrical properties of $Cr_xCe_{1-x}PO_4$ (x = 0.00, 0.08; 0.10 and 0.20) nanorods. Materials Research Bulletin. 2016;**73**:153-163

[41] Fadhalaoui A, Kouass S, Dhaouadi H. $Bi_xCe_{1-x}PO_4$ (x = 0.00, 0.02, and 0.08) nanorods: Structural, electrical, optical, and electrochemical properties. Ionics. 2018;**24**:429-450

[42] Kouass S, Fadhalaoui A, Dhaouadi H, Touati F. Electrical and electrochemical properties of undoped $CePO_4$ and doped $Ce_{0.9}Cd_{0.15-x}Li_{2x}PO_4$ nanomaterials (x = 0 and 0.02). Materials Letters. 2018;**217**:75-78

[43] Correcher V, Isasi J, Cubero A, Pérez M, Aldama I, Arévalo P, et al. Structural and luminescence characterization of synthetic Cr-doped $Ni_3(PO_4)_2$. Journal of Physics and Chemistry of Solids. 2013;**74**: 1678-1682

[44] Dhaouadi H, Fadhalaoui A, Mdani A, Rzaigui M. Structural and electrical properties of nanostructured cerium phosphate. Ionics. 2014;**20**: 857-866

[45] Jaiswal N, Kumar D, Upadhyay S, Parkash O. Effect of Mg and Sr co-doping on the electrical properties of ceria-based electrolyte materials for intermediate temperature solid oxide fuel cells. Journal of Alloys and Compounds. 2013;**577**:456-462

[46] Janene F, Dhaouadi H, Arfaoui L, Etteye N, Touati F. Nanoplate-like CuO: Hydrothermal synthesis, optical and electrochemical properties. Ionics. 2015; **21**:477-485

[47] Tian L, Zhong X, Hu W, Liu B, Li Y. Fabrication of cubic PtCu nanocages and their enhanced electrocatalytic activity towards hydrogen peroxide. Nanoscale Research Letters. 2014;**9**: 68-73

[48] Fuwei M, Dongchen W, Zhufa Z, Shumei W. Structural and electrochemical properties of $LiFe_{1-3x/2}Bi_xPO_4/C$ synthesized by sol-gel. Ionics. 2014;**20**:1665-1669

[49] Wang L, Ma P, Zhang Y, Gao C, Yan C. Determination of Li-ion diffusion coefficient *via* Coulometric titration and electrochemical impendence method. Journal of Salt Lake Research. 2009;**17**:52-55

[50] Franger S, Cras FL, Bourbon C, Rouault H. $LiFePO_4$ synthesis routes for enhanced electrochemical performance. Electrochemical and Solid-State Letters. 2002;**5**(10):A231-A233

[51] Mengyu Y, Guobin Z, Qiulong W, Xiaocong T, Kangning Z, Qinyou A, et al. In operando observation of temperature-dependent phase evolution in lithium-incorporation olivine cathode. Nano Energy. 2016;**22**:406-413

[52] Conway BE. Transition from "supercapacitor" to "battery" behavior in electrochemical energy storage. Journal of the Electrochemical Society. 1991;**138**:1539-1548

[53] Kuo SL, Wu NL. Electrochemical capacitor of $MnFe_2O_4$ with organic Li-ion electrolyte. Electrochemical and Solid-State Letters. 2007;**10**:A171-A175

[54] Huang T, Zhao C, Qiu Z, Luo J, Hu Z. Hierarchical porous $ZnMn_2O_4$ synthesized by the sucrose-assisted combustion method for high-rate supercapacitors. Ionics. 2017;**23**:139-146

[55] Xuefei D, Hailei Z, Yao L, Zijia Z, Andrzej K, Konrad Ś. Synthesis of core-shell-like ZnS/C nanocomposite as improved anode material for lithium ion batteries. Electrochimica Acta. 2017;**228**: 100-106

Chapter 5

Circuit Models of Bioelectric Impedance

Alexandru Gabriel Gheorghe, Florin Constantinescu, Miruna Nițescu and Mihai Eugen Marin

Abstract

Accurate information about fluid distribution in different compartments of the human body is very important in various areas of medicine like drug dosage, renal replacement therapy, nutritional support, coronary artery disease, colorectal cancer and HIV infection. The body impedance analysis method being simple, inexpensive, accurate and noninvasive is largely used to this end. Several models of the body impedance are presented in this chapter. The first is the Cole model, a linear, first-order *RC* circuit valid for a frequency range of two decades. Another model, developed by De Lorenzo, employs a fractional-order impedance whose parameters are identified using the frequency characteristics of the impedance module and can be used for a frequency range of three decades. In addition, two other models are presented, a ladder *RC* model valid for a frequency range of two decades and its extension to three decades, as well as a circuit containing multiple *RC* branches connected in parallel. These two models are obtained by approximating the measured body admittance modulus with a physically realizable circuit function followed by the circuit synthesis. The last model can be simplified, its simplest form being the Cole model. Allowing a better prediction of the intracellular and extracellular water volumes, this model can be viewed as an extension of the Cole model.

Keywords: bioimpedance, circuit synthesis, frequency response, impedance measurement, passive circuits

1. Overview

In various areas of medicine like drug dosage, renal replacement therapy and nutritional support, an accurate information about fluid distribution in different compartments of the body may lead to significant conclusions. Dilution methods, magnetic resonance imaging, computer axial tomography and X-ray method used to determine the fat-free mass are expensive, time consuming and unfit for routine procedures, demanding laboratories and highly trained technicians. The body impedance analysis method being simple, inexpensive, accurate and noninvasive has become largely used to predict the fluid distribution in different compartments of the body: intracellular water (ICW), extracellular water (ECW) and total body water (TBW) [1–6]. Several variants of the body impedance analysis method have been reported: single-frequency and dual-frequency bioimpedance analysis (BIA) and multi-frequency bioimpedance analysis which is also called bioimpedance spectroscopy (BIS).

Intracellular water (ICW) can be used to estimate body cell mass (BCM) which is an important indicator of the nutrition status. The evaluation of extracellular water (ECW) is also important to predict changes in fluid distribution for people suffering from wasting diseases, obesity, or patients receiving dialysis [6].

Section 2 describes the well-known Cole model, a linear, first-order *RC* circuit of the human body used for ICW and ECW volume prediction. Section 3 presents a fractional-order impedance whose parameters are identified using the frequency characteristics of the impedance module and can be used for a frequency range of up to three decades. In Section 4, two versions of a ladder *RC* model are presented, one valid for a frequency range of two decades and its extension valid for three decades. Section 5 describes a model consisting of multiple *RC* branches connected in parallel. This model can be viewed as an extension of the Cole model. Finally, some conclusions are presented in Section 6.

2. Cole model

2.1 Introduction

The frequency dependence of the body impedance, measured for example between the wrist and the ankle, can be understood knowing the behavior of the organic tissue at low frequencies (LF) and at high frequencies (HF). In the LF range (1–70 kHz), the cell membrane capacity has a high impedance value and the electric current flows mainly through ECW (**Figure 1a**). In the HF range (70 kHz–1 MHz), as this impedance decreases, the current flows through both ICW and ECW depending on their relative conductivities and volumes [7] (**Figure 1b**).

At a first glance, this behavior can be modeled with a very simple linear electrical circuit known as the Cole model [2–4] which is shown in **Figure 2**, where:

- R_i stands for the resistance of the intracellular fluid;
- C_m is the capacity of the cellular membrane; and
- R_e stands for the resistance of the extracellular fluid.

The AC equivalent resistance of this circuit at zero frequency is $R_0 = R_e$ and its AC equivalent resistance at infinite frequency is $R_\infty = (R_i \cdot R_e)/(R_i + R_e)$.

Approximating R_0 with the measured AC resistance at the minimum angular frequency ω_m and R_∞ with the measured AC resistance at the maximum angular frequency ω_M, the ICW volume V_I and the ECW volume V_E are estimated [3] as:

$$V_I = k_I \cdot Wt \cdot \left(Ht^2/R_i\right) \tag{1}$$

$$V_E = k_E \cdot Wt \cdot \left(Ht^2/R_e\right) \tag{2}$$

where Ht is the height, Wt is the weight of the subject and k_I and k_E are constants that can be determined by the cross validation against other methods [2–3].

2.2 Analysis of measurement results

Some measurement results for a bioimpedance have been reported in [4] together with a circuit model of the test bench (**Figure 3**). This circuit that includes the Cole model can be used for simulation purposes. In order to estimate the effect

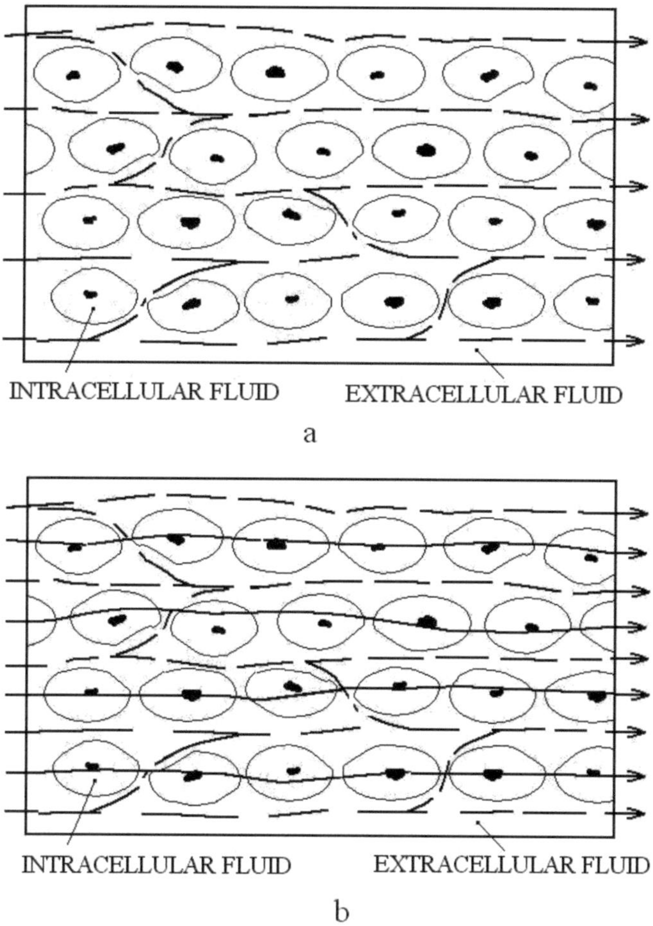

Figure 1.
The organic tissue behavior for: (a) LF electrical current flow and (b) HF electrical current flow [7].

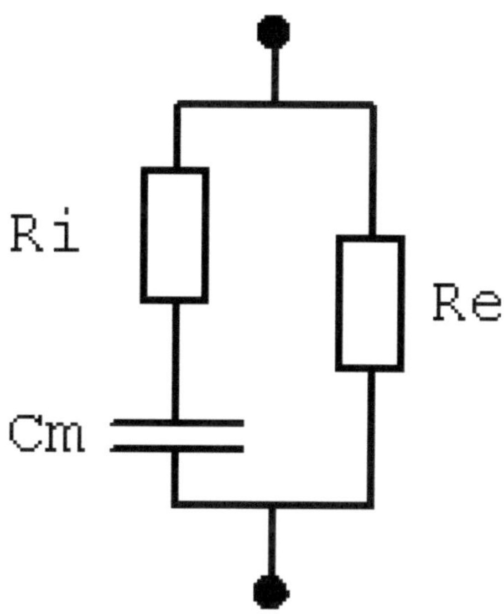

Figure 2.
Cole model.

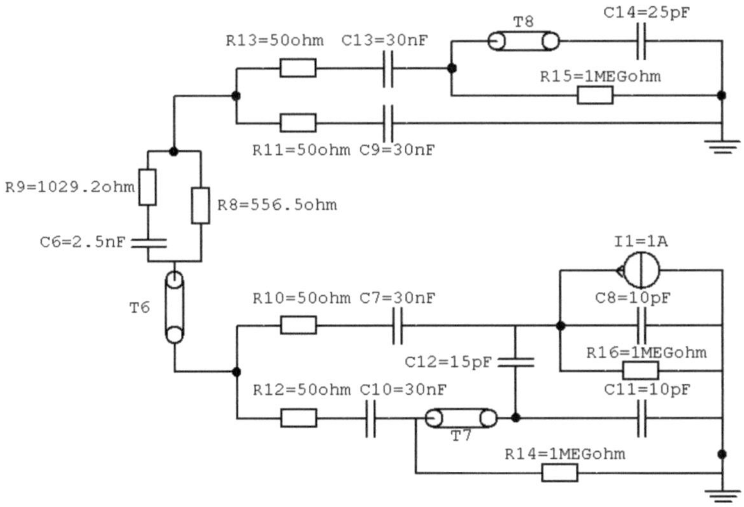

Figure 3.
The Cole model and the equivalent circuit of the measurement equipment (Model 1).

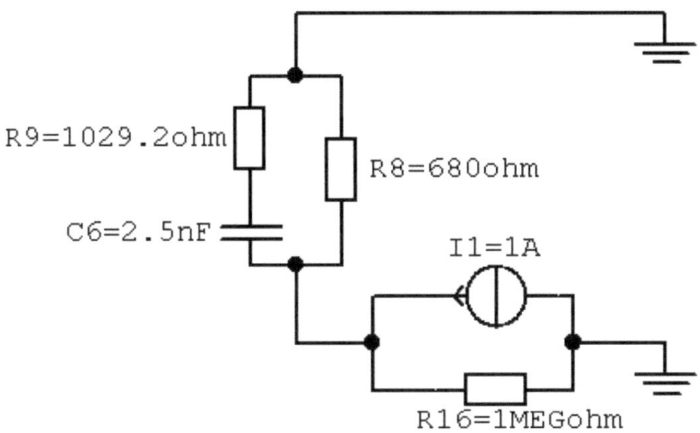

Figure 4.
The Cole model and the simplified circuit (Model 2).

of the measurement equipment (signal source, cables and connectors), the circuit in **Figure 3** has been simulated. Similar results are obtained by simulating a simpler circuit made only of the Cole model and the signal source (**Figure 4**). The frequency characteristics of the circuit which takes into account the measurement equipment (Model 1) and of the Cole model only (Model 2) are given in **Figure 5**. It follows that the measurement equipment has practically no influence and the result of the measurements is exactly the frequency characteristic of the human body bioimpedance [8].

The measured values of the human body impedance as a function of frequency are given in **Table 1** [3].

2.3 Parameter identification

The parameters of the Cole model with the frequency independent values for R_i, C_m and R_e can be identified using the measured impedance values for three

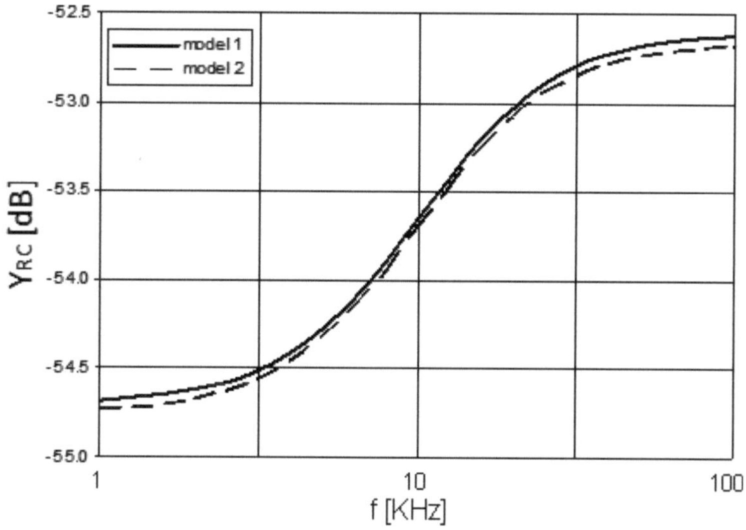

Figure 5.
Simulation results for Model 1 and Model 2.

Frequency [kHz]	Measured impedance [ohm]
1.00E+03	5.62E+02
2.00E+03	5.58E+02
3.00E+03	5.54E+02
4.00E+03	5.50E+02
5.00E+03	5.46E+02
1.00E+04	5.29E+02
1.50E+04	5.13E+02
2.00E+04	5.00E+02
2.50E+04	4.88E+02
5.00E+04	4.52E+02
7.50E+04	4.30E+02
1.00E+05	4.18E+02
1.28E+05	4.08E+02
1.48E+05	4.02E+02
1.60E+05	4.00E+02
2.00E+05	3.92E+02
2.48E+05	3.85E+02
5.00E+05	3.66E+02
7.48E+05	3.58E+02
1.00E+06	3.53E+02

Table 1.
The measured human body impedance.

frequencies. As the resistance values at the minimum and maximum frequencies are used for ECW and ICW volume estimation in Eqs. (1) and (2), we have chosen these three frequencies as $\omega_1 = \omega_m$, $\omega_2 = \omega_M$, and ω_3 corresponding to the intersection points between the measured characteristic and that of the Cole model (**Figure 6**). In this case, it is obvious that the results obtained with this Cole model are not fitted to the measured data. This proves that a more accurate model is necessary. Identifying the parameters of this Cole model in a different way, that is, using three frequencies in the middle of the frequency interval, significant errors appear at the minimum and maximum frequencies. It follows that the Cole model is not suitable for ECW and ICW computation [8].

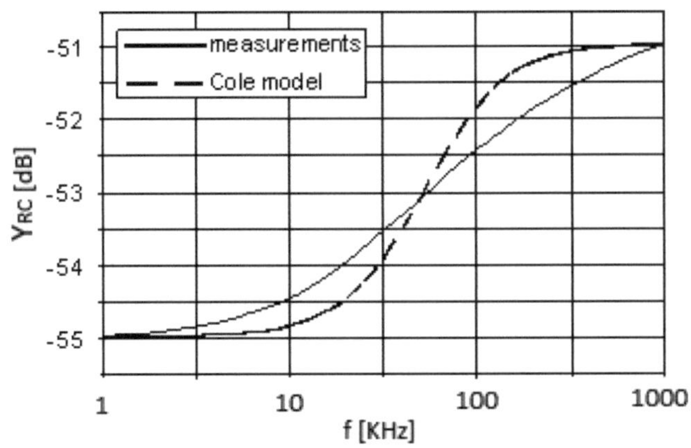

Figure 6.
The simulated Cole model and the measured results.

3. De Lorenzo model

The measurement results show that, in a wide frequency range (e.g., for two or three decades), the parameters of the circuit are frequency dependent and the relationships between the resistances of this model and the body water volumes are nonlinear. For example:

- the electrical permittivity depends on frequency as it is pointed out in **Figure** 7 [3];
- the mixture effects have a greater influence on the skeletal muscle resistivity in the LF range than in the HF range [3]; and
- due to the complexity of the nonlinear relations between R_i and R_e and ICW and ECW volumes, some heuristic relations as Eqs. (1) and Eq. (2), including the height and weight of the subject are used to compute the body water volumes.

The above properties, including unusual high values of the dielectric constant are discussed in detail in [3, 9]. As our approach is not related to these aspects, we suggest the interested researchers to read these publications.

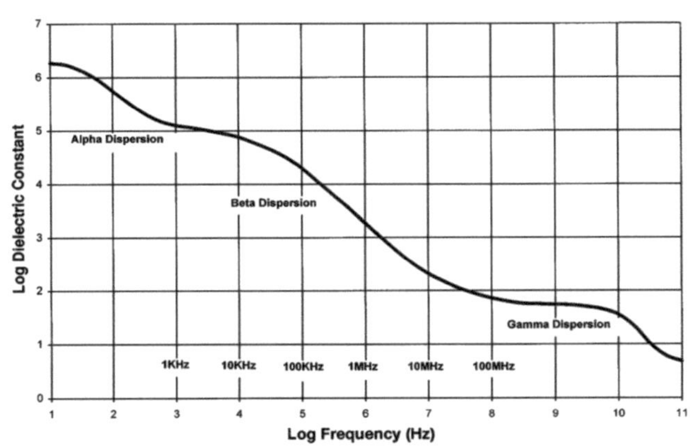

Figure 7.
Dielectric constant of the muscle tissue vs. frequency [3].

Over 500 kHz, the time delay between the excitation and its response cannot be neglected [3]. In this case, a model with distributed parameters could be more accurate.

As it was shown in the previous section, the frequency characteristic of the Cole model with frequency independent values R_i, C_m and R_e cannot be fitted to the measured data on a three-decade frequency range. In order to fix this drawback, a modified Cole model has been proposed in [3] in which the body impedance is considered as:

$$Z_{RC}(j\omega) = \left(\frac{R_e}{R_e + R_i}\right)\left(R_i + \frac{R_e}{1 + [j\omega C_m(R_e + R_i)]^{\alpha}}\right)e^{-j\omega T_d} \tag{3}$$

where ω is the angular frequency, T_d is the delay and $\alpha \in [0.3, \quad 0.7]$ is a coefficient whose value is chosen to fit the values given by Eq. (3) to the $|Z_{RC}(j\omega)|$ experimental data. As this formula does not lead to a single valued function $arg(Z_{RC}(j\omega))$, the identification of its parameters based on the measured frequency characteristics $|Z_{RC}(j\omega)|$ and $arg(Z_{RC}(j\omega))$ cannot be made. The measurements of $|Z_{RC}(j\omega)|$ and $arg(Z_{RC}(j\omega))$ suggest a circuit like that in **Figure 2** having frequency dependent components, rather than this formula that has been obtained starting from the equivalent impedance of the linear circuit in **Figure 2** in which the power $\alpha < 1$ is attached to one term while other terms remain unchanged. Taking into account that the frequency dependence of material parameters is not known for all kinds of tissues, the development of an accurate physical model is very difficult or even impossible [10].

The parameter identification for Eq. (3) is made in [3] (ignoring the uncertainty on phase) starting from measured frequency characteristics $|Z_{RC}(j\omega)|$. A very good fitting of the model characteristics to the experimental data is obtained in this case which describes de Lorenzo model. But the real parts of the body impedance in Eq. (3) at ω_m and ω_M have not the dimension of AC resistances, so the formulae Eq. (1) and Eq. (2) aimed to be employed with the Cole model AC resistances cannot be used properly.

4. *RC* ladder model

4.1 Introduction

The parameters of a model with a given structure are extracted or identified using optimization methods. In general, these methods minimize the distance between the measurement results and those obtained by simulation. In the case of semiconductor devices, the parameters of the large signal DC models or those of the small signal AC models are usually extracted using numerical techniques. Some symbolic methods have been used efficiently for parameter identification [11–13]. The circuit functions are generated using a symbolic method, obtaining analytical formulae in terms of *s* and model parameters. These parameters are computed using an optimization method to reach a global minimum of the distance between the measured and simulated values of the circuit functions for a set of test frequencies. The symbolic methods are very efficient for the computation of derivatives which are usually needed in the optimization procedure. The optimization can be performed using genetic algorithms [14]. Sometimes, hierarchical techniques are employed to obtain combined DC-AC models [15].

The body impedance analysis method has not yet reached its full potential. Following the trend to improve the method by increasing the level of model accuracy, a new approach to the parameter identification for a linear *RC* model in bioimpedance spectroscopy is presented in this section. This approach employs the approximation of the measured body admittance modulus $|Y_{RC}(j\omega)|$ with a physically realizable function followed by the circuit synthesis [16]. This model is a linear *RC* circuit with frequency independent values of resistances and capacitances. As the frequency dependence of the phase angle $arg(Y_{RC}(j\omega))$ can be computed from $|Y_{RC}(j\omega)|$ using the Bayard-Bode relationships [17], the measured values of $arg(Y_{RC}(j\omega))$ are not needed for the parameter identification of this model. Two equivalent circuits of the human body, built using this approach, have been proposed [10–11]. These are ladder circuits which cannot be considered as extensions of the Cole model.

4.2 *RC* admittance synthesis

The synthesis method for an *RC* admittance (Y_{RC}) developed in [15] can be used also for an *RC* impedance ($Z_{RC} = 1/Y_{RC}$) with some minor modifications, and it is presented in the following.

A function $F(s)$ of a complex variable s is an *RC* admittance if and only if the following conditions are fulfilled:

- $F(s)$ is a rational fraction of s with real coefficients;
- the poles and zeros of $F(s)$ are simple and alternate on the negative real axis, the closest to the origin being a zero; and
- the number of zeros of $F(s)$ is equal or greater with one with respect to the number of poles $F(s)$.

Replacing s with $j\omega$, (where $\omega=2\pi f$ and f is the frequency) the shape of the *RC* admittance modulus curve $|Y_{RC}(j\omega)|$ versus ω is defined by the poles and zeros location. Sweeping the ω axis starting from the origin, it can be observed that the location of a zero is associated with a slope change of 20 dB/decade and the location of a pole is associated with a slope change of −20 dB/decade. This is because the characteristic $|Y_{RC}(j\omega)|$ has asymptotes whose slopes are 20 dB/decade, 0, 20 dB/decade, 0, and so on. The $|Y_{RC}(j\omega)|$ characteristic approximation by asymptotes has the maximum error of 3 dB at the asymptote intersection (**Figure 8**) [16].

A natural way to approximate the $|Y_{RC}(j\omega)|$ characteristic is to consider a smaller asymptote number than that corresponding to the measured characteristic. A greater error ε between measured and simulated values leads to a simpler circuit (**Figure 8**).

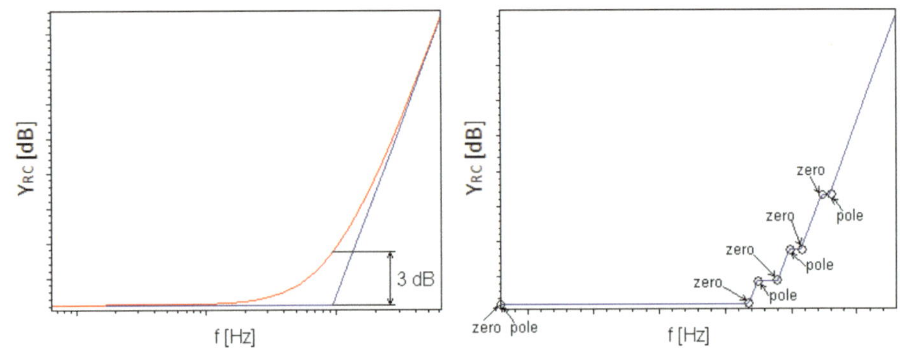

Figure 8.
The $|Y_{RC}(j\omega)|$ *characteristic approximation by asymptotes.*

The algorithm for the synthesis of a *RC* one-port in the angular frequency band $[\omega_m, \omega_M]$, where ω_m is the minimum value and ω_M is the maximum value, has the following steps [16]:

- set the first zero z_1 corresponding to the minimum angular frequency ω_m;
- compute the remaining poles and zeros at slope changes, by sweeping the ω axis with the step $\Delta\omega$; a larger $\Delta\omega$ leads to a simpler circuit;
- compute the *RC* admittance expression; and
- compute the circuit parameters using a synthesis method.

Sweeping the frequency axis with a step $\Delta\omega_m$, the algorithm checks the error between the 20 dB/decade asymptote and the given characteristic. This error cannot be greater than an imposed value ε. The first pole p_1 is assigned to the last value before that corresponding to an error of 2ε or greater. If this error occurs after the first angular frequency step $\Delta\omega_m$, then p_1 is placed in the vicinity of z_1. Afterwards, the first asymptote is translated so that a maximum error of ε is obtained. The other asymptotes are determined similarly, in order to fulfill the condition $error \leq \varepsilon$ for each asymptote.

4.3 Parameter identification for the two-decade model

The following models are build starting from measurement of the body impedance frequency characteristic $|Z_{RC}(j\omega)|$ reported in [3, 8]. The simulations presented in Section 2 show that the result of the frequency characteristic measurement is an accurate representation of the human body impedance modulus. Using the above algorithm it follows:

$$Y_{RC}(s) = \frac{17.81 \cdot 10^{-4}\left(15.92 \cdot 10^{-5}s + 1\right)\left(10.23 \cdot 10^{-6}s + 1\right)\left(26.82 \cdot 10^{-7}s + 1\right)\left(28.82 \cdot 10^{-8}s + 1\right)}{\left(15.91 \cdot 10^{-5}s + 1\right)\left(86.25 \cdot 10^{-7}s + 1\right)\left(22.70 \cdot 10^{-7}s + 1\right)} \tag{4}$$

Using a Cauer synthesis, the continued fraction expansion in Eq. (5) is obtained, and the circuit is given in **Figure 9**. The parameter values extracted from Eq. (5) are given below.

$$Y_{RC}(s) = 71.97 \cdot 10^{-11}s + \cfrac{1}{410.65 + \cfrac{1}{31.08 \cdot 10^{-9}s + \cfrac{1}{106.17 + \cfrac{1}{18.16 \cdot 10^{-8}s + \cfrac{1}{44.66 + \cfrac{1}{15.69 \cdot 10^{-3}s + 98.64}}}}}} \tag{5}$$

A comparison between the frequency characteristic of the *RC* ladder model, the Cole model and the measured results is given in **Figure 10**.

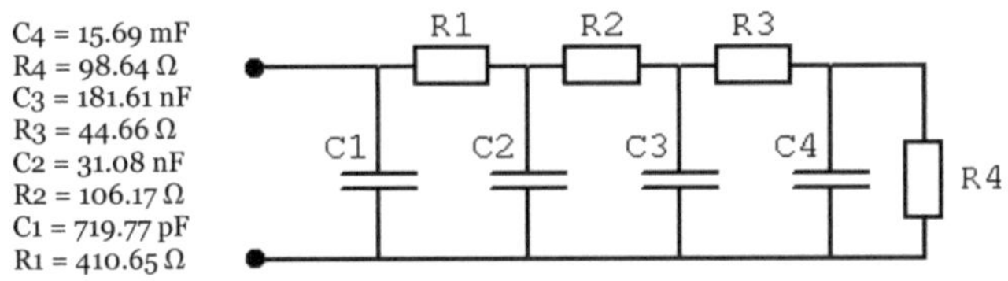

Figure 9.
Cauer synthesis of the RC *ladder circuit model [8].*

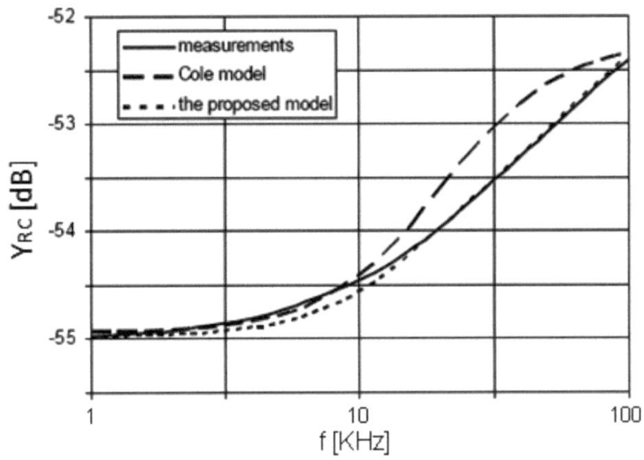

Figure 10.
The frequency characteristic for the Cole model, the two-decade RC *ladder model and the measured results.*

4.4 Parameter identification for the three-decade model

Using the same algorithm presented above applied for the measurement set of three decades, a more elaborated admittance expression, given in Eq. (6), is obtained [18]. The Cauer synthesis starting from the continued fraction expansion in Eq. (7) gives the circuit presented in **Figure 11**. The parameter values are given above the model. A comparison between the frequency characteristic of the new model and the measured results is given in **Figure 12**.

$$Y_{RC}(s) = \frac{17.81 \cdot 10^{-4}\left(15.92 \cdot 10^{-5}s + 1\right)\left(13.04 \cdot 10^{-6}s + 1\right)\left(43.11 \cdot 10^{-7}s + 1\right)\left(18.59 \cdot 10^{-7}s + 1\right)\left(75.82 \cdot 10^{-9}s + 1\right)}{\left(15.91 \cdot 10^{-5}s + 1\right)\left(11.55 \cdot 10^{-6}s + 1\right)\left(38.08 \cdot 10^{-7}s + 1\right)\left(16.47 \cdot 10^{-7}s + 1\right)} \tag{6}$$

$$Y_{RC}(s) = 19.47 \cdot 10^{-11}s + \cfrac{1}{392.53 + \cfrac{1}{23.94 \cdot 10^{-9}s + \cfrac{1}{109.02 + \cfrac{1}{10.71 \cdot 10^{-8}s + \cfrac{1}{42.54 + \cfrac{1}{50.07 \cdot 10^{-8}s + \cfrac{1}{17.39 + \cfrac{1}{16.95 \cdot 10^{-3}s + 106.8}}}}}}}} \tag{7}$$

C5 = 16.99 mF, R5 = 106.80 Ω, C4 = 500.71 nF, R4 = 17.39 Ω, C3 = 107.12 nF, R3 = 42.54 Ω, C2 = 23.94 nF, R2 = 109.02 Ω, C1 = 194.77 pF, R1 = 392.53 Ω.

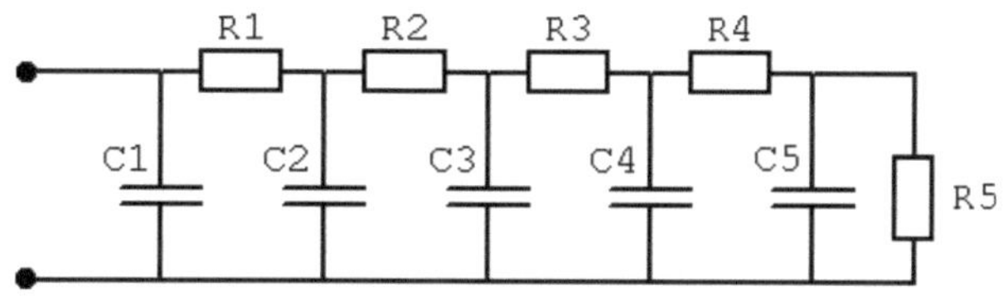

Figure 11.
The Cauer synthesis of the three-decade RC *ladder model.*

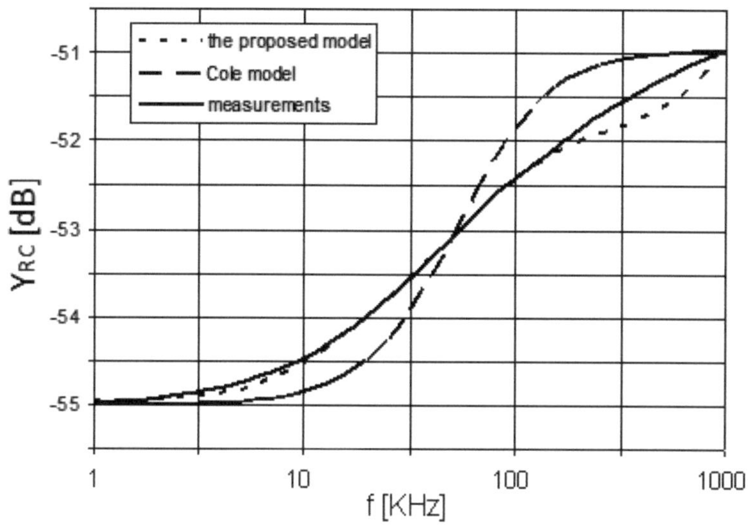

Figure 12.
The frequency characteristic of the Cole model, the three-decade RC *ladder model and the measured results.*

5. *RC* parallel model

5.1 Introduction

A behavioral model, as a linear circuit which can be an extension of the Cole model is the best choice, taking into account that the intracellular and the extracellular water volumes are related to the real part of the model impedance computed at minimum and maximum frequencies [3], this impedance being well defined only for a model of this kind.

An *RC* parallel model, valid for a frequency range of three decades, which can be reduced to the Cole model for a narrow frequency interval, is presented in this section.

5.2 Parameter identification for the *RC* parallel model

For the parameter identification of the *RC* parallel model, only the measured frequency characteristic $|Y_{RC}(j\omega)|$ is used [10]. In order to build this model, the approximation method is employed, followed by the circuit synthesis as it is described in the previous section.

Using the above algorithm, the frequency characteristic $|Y_{RC}(j\omega)|$ corresponding to the data in [3] has been approximated by the admittance in Eq. (6) with an error ε = 0.95 dB using a sweeping step $\Delta\omega_m$ = 8315 Hz. The synthesis of this admittance can be made by the Foster II method which gives the most interesting circuit in **Figure 13**.

The direct employment of the Foster II synthesis algorithm starting from Eq. (6) leads to some negative parameter values. This effect can be avoided performing the Foster II synthesis of $|Y_{LC}(j\omega)|$, where $Y_{LC}(s)$ in Eq. (9) is given by the frequency transformation in Eq. (8) [17].

$$Y_{LC}(s) = \frac{1}{s} \cdot Y_{RC}(s^2) \tag{8}$$

$$\begin{aligned} Y_{LC}(s) = 19.47 \cdot 10^{-11} s + \frac{17.81 \cdot 10^4}{s} + \frac{24.14 \cdot 10^{-10} s}{11.54 \cdot 10^{-6} s^2 + 1} + \\ + \frac{94.53 \cdot 10^{-11} s}{38.08 \cdot 10^{-7} s^2 + 1} + \frac{50.94 \cdot 10^{-11} s}{16.47 \cdot 10^{-7} s^2 + 1} + \frac{59.04 \cdot 10^{-13} s}{15.91 \cdot 10^{-5} s^2 + 1} \end{aligned} \tag{9}$$

Starting from the partial fraction decomposition in Eq. (9), the parameter values are: C5 = 5.9 pF, R5 = 27 MΩ, C4 = 2.41 nF, R4 = 4.78 kΩ, C3 = 0.945 nF, R3 = 4.03 kΩ, C2 = 0.51 nF, R2 = 3.23 kΩ, C1 = 0.195 nF, R1 = 561.5 Ω.

The resistance corresponding to the volume of the extracellular water can be computed for f_{min} = 1 kHz and has a 560.97 Ω value, which is practically the same with R_E = 562 Ω given by the Cole model.

The resistance corresponding to the volume of the intracellular water can be computed for f_{max} = 1000 kHz and has a 314.97 Ω value, unlike R_i = 352.69 Ω given by the Cole model. Due to the better agreement with experimental data, it is expected that the body water volume prediction will be improved considering these values in Eqs. (1) and (2).

It is very interesting to observe that R1 has a similar value to R_e in the Cole model, being the equivalent resistance for f = 0 Hz. This circuit can be viewed as a generalization of the Cole model. The two branch models contain R1, R2, C2. As the frequency range of interest is extended to higher frequencies, a model with a greater number of branches is needed. The simulated data obtained with models with various numbers of branches, obtained by imposing the same error ε on various frequency intervals are given in **Figure 14**.

A similar circuit (**Figure 15**) is given in [3] without pointing out how the resistance and capacitance values can be computed starting from the measured data.

In order to appreciate the agreement between the measured and simulated data, the measuring errors must be known. Unfortunately, no information on these errors is given in [3].

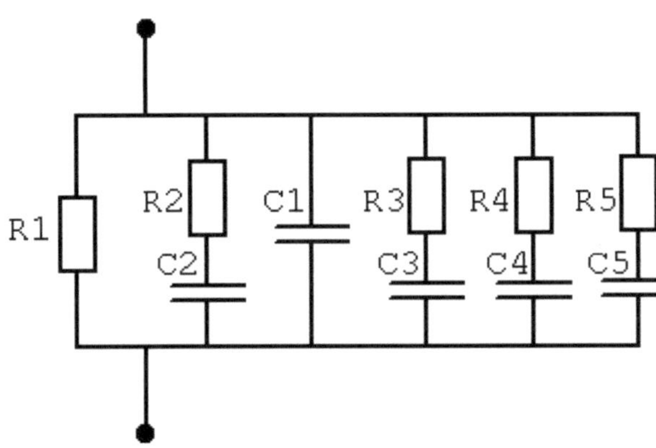

Figure 13.
Foster II synthesis of the new circuit model [10].

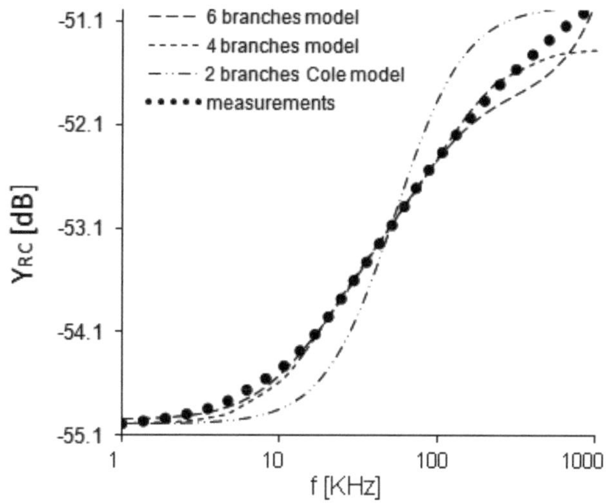

Figure 14.
The measured frequency characteristic and some proposed models with 2, 4 and 6 branches [10].

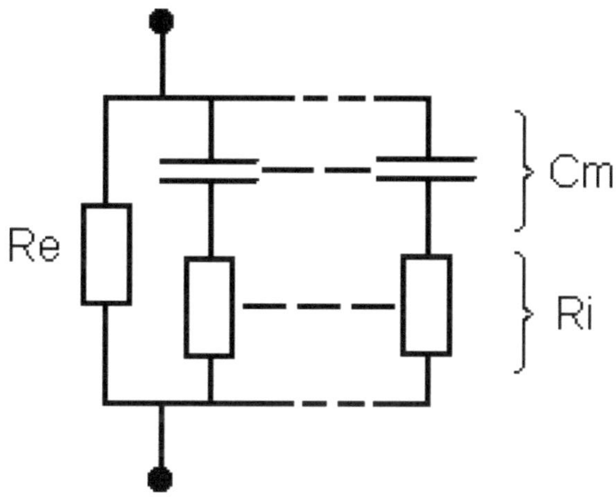

Figure 15.
The extended Cole model [3].

6. Conclusions

Four models of the human body bioimpedance used to compute ICW and ECW volumes have been presented in this chapter.

The first model presented in this chapter is the Cole model. This model is used for body water volume prediction, having frequency independent values for R_i, R_e and C_m. It cannot reproduce the measurement results for a three or even for a two-decade frequency range.

The second model is based on a fractional exponent formula for the body impedance whose module is fitted to the measured values in [3]. But the real part of this impedance at the minimum and maximum frequencies cannot be computed and, consequently, the ICW and ECW formulae, having an outstanding practical importance, cannot be used.

The next two behavioral models are based on parameter identification. These models are linear *RC* circuits with frequency independent elements, whose parameters can be identified starting from the measured values $|Z_{RC}(j\omega)|$ reported in [3]. The influence of the measurement equipment including signal source, cables (modeled as transmission lines) and connectors has been shown to be negligible, so $|Z_{RC}(j\omega)|$ given in [3] is an accurate representation of the human body impedance modulus [8, 18].

After the synthesis of the third model, an *RC* ladder, valid for a frequency range between 1 and 100 kHz [8], and its extension to a three-decade frequency interval [17], the fourth model, the *RC* parallel circuit [9], whose validity range is three decades is presented. This model contains some *RC* branches connected in parallel. This model can be simplified, taking into account that the influence of some branches is negligible in a certain frequency range, its ultimate simplification being the linear *RC* Cole model. It follows that this model can be considered as an extension of the linear *RC* Cole model, allowing a good prediction of the intracellular and extracellular water volumes. All these linear lumped *RC* circuits avoid using both intricate frequency dependent elements suggested by the physical interpretation of current conduction in human body and the fractional exponent impedance formula of de Lorenzo model [3].

Even though the modeling of fractional-order circuits is rigorously established [19], a linear *RC* circuit model based on straightforward concepts is more useful for intracellular and extracellular water volume prediction than a fractional-order system. The development of these new models illustrates the actual trend [20] to make noninvasive investigation methods more precise in various areas of medicine [7] as coronary artery disease [21], colorectal cancer [22] and HIV infection [23].

Author details

Alexandru Gabriel Gheorghe*, Florin Constantinescu, Miruna Nițescu and Mihai Eugen Marin
University Politehnica of Bucharest, Romania

*Address all correspondence to: alexandru.gheorghe@upb.ro

References

[1] Mattie J, Zarowitz B, De Lorenzo A, Andreoli A, Katzarski K, Pan G, et al. Analytic assessment of the various bioimpedance methods used to estimate body water. Journal of Applied Physiology. 1998;**84**:1801-1816

[2] Lichtenbelt WDM, Westerterp KR, Wouters L, Luijendijk SCM. Validation of bioelectrical-impedance measurements as a method to estimate body-water compartements. American Journal of Clinical Nutrition. 1994;**60**: 159-166

[3] De Lorenzo A, Andreoli A, Matthie J, Withers P. Predicting body cell mass with bioimpedance by using theoretical methods: A technological review. Journal of Applied Physiology. 1997; **82**(5):1542-1558

[4] Gudivaka R, Schoeller DA, Kushner RF, Bolt MJG. Single- and multifrequency models for bioelectrical impedance analysis of body water compartments. Journal of Applied Physiology. 1999;**87**(3):1087-1096

[5] Matthie JR. Second generation mixture theory equation for estimating intracellular water using bioimpedance spectroscopy. Journal of Applied Physiology. 2005;**99**(2):780-781

[6] Earthman C, Traughber D, Dobratz J, Howell W. Bioimpedance spectroscopy for clinical assessment of fluid distribution and body cell mass. Nutrition in Clinical Practice. 2007; **22**(4):389-405

[7] Marin CV. Frequency selection for parameter identification in bioimpedance spectroscopy. Revue Roumaine Des Sciences Techniques. 2009;**54**(4):425-434

[8] Gheorghe AG, Marin CV, Constantinescu F, Nitescu M. Synthesis of a new RC model for body cell mass prediction. In: National Symposium on Theoretical Electrical Engineering, Politehnica University, June 5–7. 2008

[9] Schwan HP. Electrical properties of tissues and cell suspensions: Mechanisms and models. In: Proceedings of 16th Annual International Conference of the IEEE Engineering in Medicine and Biology Society. 1994. pp. 70-71

[10] Gheorghe AG, Marin CV, Constantinescu F, Nitescu M. Parameter identification for a new circuit model aimed to predict body water volume. Advances in Electrical and Computer Engineering. 2012;**12**(4):83-86

[11] Constantinescu F, Marin CV, Nitescu M, Marin D. Parameter identification using symbolic pole/zero expressions. In: European Conference on Circuit Theory and Design (ECCTD'03), Poland, September 1–4. 2003

[12] Avitabile G, Fedi G, Giomi R, Luchetta A, Manetti S, Piccirilli MC. Parameter extraction in electronic device modelling using symbolic techniques. In: Proceedings of the Second International Workshop on Symbolic Methods and Applications to Circuit Design, October 8–9. Germany: Kaiserslautern; 1998. pp. 253-261

[13] Konkzykowska A, Rozes P, Bon M. Parameter extraction of semiconductor devices electrical models using symbolic approach. In: Proceedings of the Second International Workshop on Symbolic Methods and Applications to Circuit Design, October 8–9. Italia: Firenze; 1992. pp. 1-10

[14] Burmen A, Tuma T. Model parameter identification with SPICE OPUS: A comparison of direct search and elitistic genetic algorithm. In: Proceedings of ECCTD'01. 2001. pp. 61-64

[15] Bandler JW, Chen SH, Ye S, Zhang Q-J. Integrated model parameter extraction using large scale optimization concepts. IEEE Transactions on Microwave Theory and Techniques. 1988;**36**(12):1629-1638

[16] Constantinescu F, Gheorghe AG, Ioan CD, Nitescu M, Iordache M, Dumitriu L. A new approach to the computation of reduced order models for one-port and two-port RC circuits. In: International Symposium on Circuits and Systems (ISCAS), May 21–24. Greece: Island of Kos; 2006. pp. 4002-4005

[17] Guillemin EA. Synthesis of Passive Networks - Theory and Methods Appropriate to the Realization and Approximation Problems. New York: John Wiley & Sons; 1967

[18] Gheorghe AG, Marin CV, Constantinescu F, Nitescu M. A new circuit model for body cell mass prediction. In: 4th European Conference on Circuits and Systems for Communications (ECCSC 08), Jul 10–11. Romania: Politehnica University, Bucharest; 2008

[19] Elwakil AS. Fractional order circuits and systems: An emerging interdisciplinary research area. IEEE Circuits and Systems Magazine. 2010; **10**(4):41-44

[20] Zhurbenko V. Challenges in the design of microwave imaging systems for breast cancer detection. Advances in Electrical and Computer Engineering. 2011;**11**(1):91-96

[21] Singh RB, Niaz MA, Beegom R, Wander GS, Thakur AS, Rissam HS. Body fat percent by bioelectrical impedance analysis and risk of coronary artery disease among urban men with low rates of obesity: The indian paradox. Journal of the American College of Nutrition. 1999;**18**(3):268-273

[22] Gupta D, Lammersfeld CA, Burrows JL, Dahlk SL, Vashi PG, Grutsch JF, et al. Bioelectrical impedance phase angle in clinical practice: Implications for prognosis in advanced colorectal cancer. The American Journal of Clinical Nutrition. 2004;**80**:1634-1638

[23] Earthman CP, Mattie JR, Reid PM, Harper IT, Ravussin E, Howell WH. A comparison of bioimpedance methods for detection of body cell mass change in HIV infection. Journal of Applied Physiology. 2000;**88**:944-956

Chapter 6

Electrodermal Activity: Simultaneous Recordings

Haval Y. Yacoob Aldosky and Dindar S. Bari

Abstract

Electrodermal activity (EDA) is a sensitive measure of the sympathetic nervous system activity. It is used to describe changes in the skin electrical properties. This chapter aimed to show advantages of simultaneous recordings of EDA parameters at the same skin site over other recordings. The literature databases, Web of Science and Google Scholar, were searched using terms like "electrodermal activity," "sequential recording," "simultaneous recording," "skin conductance," "skin potential," and "skin susceptance." Articles that include sequential and/or simultaneous recording of EDA parameters were analyzed. The chapter presents a description of the oldest and current methods used for recording EDA parameters and an explanation of the newest techniques used in EDA researches. Although sequential recordings are predominant and widely spreading, much effort has been made to simultaneously record skin conductance (SC) and skin potential (SP), and recently researchers realized the capability of simultaneously recording SC, SP, and skin susceptance (SS) at the same skin site. The advantage of simultaneous over the sequence measurements is that the latter must be manually time realigned when measured by different instruments, which means it is time-consuming. Although the simultaneous measurements are used exclusively for research purposes at this stage, this may open horizons in the modern trends of psychophysiology applications in the near future.

Keywords: electrodermal activity (EDA), simultaneous SC and SP, skin conductance, skin potential, skin susceptance, EDA parameters

1. Introduction

Numerous studies had been done by researchers to analyze the basic of EDA responses as indicators via employing various methods and techniques of measurement since long time ago. However, results and hypotheses of several such studies were rather conflicting. These investigations were apparently lost of sight by later investigators, and many of the same errors and conflicting results have appeared in later studies. The purpose of this review was to advance the understanding of simultaneous recordings of EDA parameters at the same skin site and show its advantages over sequential or alternate recordings.

1.1 Skin

To reach our goal in this chapter, a brief introduction about the skin, which is one of the most complex organs of the human body, should be depicted.

The human skin is a complex and a large organ (in terms of both weight and surface area) that covers the body and forms a remarkable protective barrier against the external environment [1, 2]. It is facilitating to regulate the core body temperature and water balance via bloodstream to the exterior of the body [3].

EDA responses are frequently used as the peripheral indicators of sympathetic activation. The EDA measurement by psychophysiologists is basically concerned with sweat gland activity that is psychologically induced. Different internal and external stimuli cause mental stress; as a result, sweat glands produce various amounts of sweat that are propelled up to the sweat ducts and hence result in different EDA responses. Numerous models have been proposed to explain how these peripheral mechanisms are associated to the electrical activity of the skin and to the transient increases in EDA parameters evoked by external stimuli. According to Edelberg [4] one can account for the several EDA phenomena, including alteration in tonic skin conductance level (SCL) and phasic skin conductance response (SCR) amplitude, with a model based completely on the sweat glands [4]. As noted by Edelberg [5], one should not be surprise that an organ with such vital and dynamic functions continuously receives signals from control centers in the brain, and the author suggested that "we can listen in on such signals by taking advantage of the fact that their arrival at the skin is heralded by measurable electrical changes that we call electrodermal activity" (p. 368).

In order to clearly understand how EDA is linked to the sweat glands, it is useful to imagine the sweat ducts as a set of variable resistors with parallel connection. Sweat columns will rise in the ducts with different amounts and different sweat gland numbers, depending on the level of the sympathetic nervous system activation. As the ducts are filled through sweating, there is a more conductive path through the relatively resistant stratum corneum layer. As the sweat level further rises, the resistance in that variable resistor is further lowered. Changes in the sweat level in the ducts alter the values of the variable resistors and thus yield observable changes in EDA [3].

1.2 Electrical bioimpedance

Electrical bioimpedance is a measure of how well the biological tissues such as the skin impede alternating current flow at different frequencies. Electrical impedance has two components: the resistive and the reactive parts. Mathematically, the electrical impedance (Z) is expressed as a complex number by the sum of the resistance (R) and the reactance (X):

$$Z = R + jX \tag{1}$$

Electrical impedance is the ratio between the voltage and current. When a known current is applied to a material, the impedance is found by measuring the voltage between the electrodes and dividing it by the current. However, in many cases, applying a known voltage to the material and measuring the resulting current between the electrodes are more practical. The measured current then becomes inversely proportional to the impedance. This quantity is called electrical admittance (Y), which allow current to flow. It is also expressed as a complex number with two components: conductance (G) and susceptance (B):

$$Y = 1/Z = G + jB \tag{2}$$

Both the X and B are dependent on the frequency (F) of the applied current through the material and only can be measured by alternating the direction of the current. B is proportional to this frequency, and the electrical capacitance (C) of the material can be calculated from Eq. (3).

$$B = 2\Pi \, FC \tag{3}$$

1.3 Electrical impedance spectroscopy

Electrical impedance spectroscopy (EIS) is applying a sinusoidal voltage or current to the sample under test to calculate impedance parameters within a wide range of frequencies, where frequency-dependent electrical properties of biological tissues can be detected. Therefore, EIS has been proven as an effective technique for noninvasive tissue characterization in medical, biomedical, and biological applications [6]. The real part of the impedance is associated with resistive pathways across the tissues, which is typically large at low frequencies but decreases with increasing frequency, whereas the imaginary part of the impedance is associated with capacitive pathways, which decreased (not noticeable) at high frequencies [7].

1.4 Electrodermal activity (EDA)

EDA is the preferred term for changes in electrical properties of the skin. It is a set of physiological parameters of sympathetic nervous system activity, and it has been used for physiological measurements due to a strong link with the autonomous activity [8]. However, the EDA phenomenon and its appearance are not sufficiently clarified yet [9, 10]. EDA is measured from the eccrine glands, which cover most of the body. In addition, they are concentrated in the palmar and plantar dermatomes, and, therefore, these are known to be the best sites for measuring EDA [11, 12]. Mainly, there are two categories of electrodermal recordings, namely, endosomatic and exosomatic measurements. In endosomatic measurements, only potential differences originating in the skin itself are recorded without using any external source of current. In exosomatic measurements, externally very small amount of current [either alternating current (AC) or direct current (DC)] is applied to the skin. This is frequently used to measure SC, and in some recently published studies, it is also used to measure skin susceptance. EDA signals are a manifestation of the eccrine sweat gland activity that is innervated by the autonomic nervous system, primary by the sudomotor nerves [13]. When the sudomotor nerves stimulate the sweat production, indeed the SC changes as a result of sweat secretion and alterations in ionic permeability of sweat gland membranes [5, 11, 14].

EDA is composed of two basic components (**Figure 1**): tonic (level) and phasic (response), each with various time scales and relationships with the stimuli. Tonic EDA is represented by SCL which represents the slow-changing baseline level of the SC and skin potential level (SPL) which represents the slow-varying baseline level of the SP. Alterations in the SCL are thought to reflect slow changes in the autonomic nervous system dynamics. Phasic EDA is specified by a fast varying component, known as the SCR and skin potential response (SPR). Both EDA phenomena, tonic (SCL and SPL) and phasic (SCR and SPR), are generated under autonomous nerve control of the active organs of the skin [15], which is reflecting the elicited response of the eccrine sweat glands to external stimuli [11, 14]. Some recent evidences suggest that these two components depend on various neural mechanisms [16] and, consequently, both carry relevant and non-redundant information about the autonomic nervous system dynamic activity [14]. EDA is employed in a broad range of experimental setups since it is a relatively straightforward measurement providing valuable information on the autonomic nervous system response to a wide range of externally applied stimuli. Particularly, SC

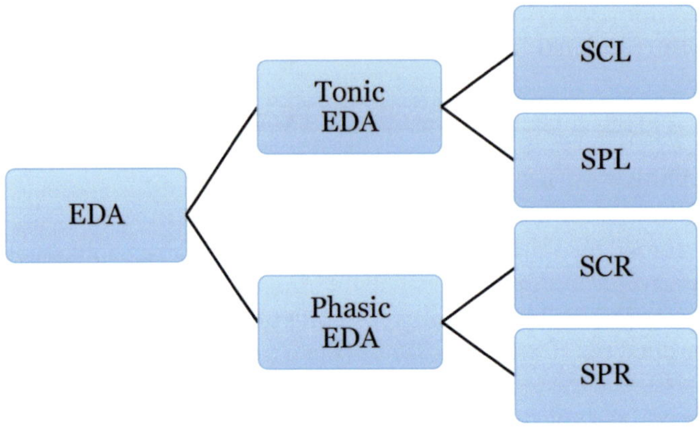

Figure 1.
Basic components of EDA.

analysis is commonly used to quantify the levels of arousal related with cognitive and emotional processes [3, 14, 17, 18].

In spite of sweating being basically a means of thermoregulation, sweat glands located on the plantar and palmar surfaces perhaps evolved to increase grip and enhance sensitivity, possibly more responsive to psychologically stimuli than to thermal stimuli [3, 5, 11, 14]. This relationship between EDA, psychological stimuli, and autonomic nervous system makes this physiological signal broadly popular in neuroscience research studies, including quantification of arousal levels during cognitive and emotional processes, information processing, and clinical research examining predictors and correlates of normal and pathological behavior [3, 17–19], such as psychopathology, personality disorders, and neuropsychology [14].

2. Historical background of EDA

The study of the electrical changes and properties of the human skin began over 100 years ago [3]. Since then, various terms have been utilized to describe this phenomenon including psychogalvanic skin response, skin resistance response, skin conductance response, skin potential response, and EDA. Historically, the most widely used term to label this phenomenon was the galvanic skin response (GSR). However, in 1966, EDA was first introduced by Johnson and Lubin as a common term for all electrical phenomena in the skin, including all passive and active electrical properties which can be traced back to the skin and its appendages [11].

Neumann and Blanton [20] thoroughly reviewed the history of EDA research, which dates back to experiments performed in 1849 by du Bois-Reymond in Germany. In his experiments, du Bois-Reymond used a zinc sulfate solution to put either hands or feet in it and consequently observed that an electrical current going from the limb at rest to the other one was contracted voluntarily. However, du Bois-Reymond considered that the observed phenomenon was due to muscle action potentials [11].

Hermann and Luchsinger from Switzerland in 1878 were the first to experimentally show a connection between sweat gland activity and flow of current in the skin. They observed that an electrical stimulation of the sciatic nerve in the curarized cat caused sweat secretion and an electric current in the footpad on the same side of body [11]. Three years later, the voluntary movement experiment that was performed by du Bois-Reymond more than 30 years ago was repeated by Hermann. It was found that palms and fingers showed greater skin current than other sites of the body such as the wrist and elbow areas due to the greater concentration of sweat glands and therefore stronger sweating in that regions [20]. In 1879 Vigouroux was the first who observed that psychological factors are related to EDA. He measured changes in skin resistance (SR) that changed in parallel with the amount of anesthesia in hysterical patients. In addition, he presumed that both phenomena were dependent upon central processes, but he did not believe that the swift changes in SR observed by himself could be resulted by local processes in the skin itself [11].

The pioneering studies of electrodermal phenomena, however, have been done by Fere (1888) and Tarchanoff (1889) [3, 11]. Fere by employing an external direct current observed reduction in SR following emotional stimulation in hysterical patients [11]. Tarchanoff could measure changes in skin electrical potential between two electrodes connected to the skin surface without the aid of any external electromotive force. He supposed that the EDA phenomena observed by him were due to sweat gland activity, which is dependent on the secretory nerve action. He observed that, even at rest, the current flow from regions of high concentration in sweat glands to those poor in them. In contrast, Fere assumed that decrease in SR following stimulation was due to a decrease in blood flow of the skin, i.e., resulted from partial displacement of the blood peripheral resistance by the lower interstitial fluidity resistance [3, 11].

In 1928 and 1929, a decisive contribution to the investigation of the origins of endosomatic EDA was made by Gildemeister and Rein. They for the first time restricted the locus of SP origin to only one of the two recording sites by injuring the skin under the other electrode, where no SP of its own could develop [11]. In 1930 the sweat glands were identified as the seat of the "psychogalvanic phenomena" by McClendon and Hemingway. In the same year (1930), a palmar galvanic test was used for indicating sweat secretion by Wang and Lu. Later EDA measurements became a common mechanism within the field of psychophysiology [19, 21].

SC units were strongly supported by Darrow [22] as well as Lykken and Venables [23] as being sufficient with respect to physiological models of the peripheral mechanisms of EDA. An electrical model of the skin was proposed by Edelberg [24] after having performed EDA research for more than 10 years, which takes into account the existence of polarization capacitances [11]. Using this background, psychophysiological aspects of several EDA components in details were first established by Edelberg [5] including parameters which were subsequently focused on [11].

Nowadays EDA measurement is regarded as "one of the most widely used response systems in the history of psychophysiology" [3] (p. 159). In addition, over the last decades, the areas for application of EDA have been steadily widened such as in the field of engineering psychophysiology as well as in neurology [11]. As EDA is a generic reference that subsumes all methods of measuring the electrical activity of the skin, it is the preferred term to label this phenomenon [25].

3. EDA recording systems

3.1 Sequentially recording SC and SP

Some authors tried to measure EDA parameters at the same skin site, but such parameters were sequentially recorded.

Venables and Sayer [26] had determined the relationship between the SC and SP, by means of two measuring systems. The idea was to record both EDA parameters at the same skin (palm and inner surface of the left arm), but not simultaneously. They stated that due to the disturbance in the measurement of SP created by the presence of the impressed voltage used to measure SC, the two measures were not taken simultaneously. It was found that the two measures are related, although not so highly as to make them equivalent [26].

The role of sweat gland activity in the mediation of SC and SP levels was investigated by Venables and Martin [27]. Both SP and SC were recorded at the same skin site (tips of the fore, middle, and ring fingers of each hand), but not simultaneously [27].

Turpin and Siddle [28] presented effects of series of auditory stimuli on the sequential SCR and SPR recordings. Bipolar recording of SC was accomplished by attaching electrodes on the medial phalanx of the index and second fingers of the non-preferred hand of the subject. Unipolar recording of SP was made by placing electrodes on the medial phalanx of the third finger of non-preferred as the active site, while the reference site was a point two thirds of the distance from the wrist to the elbow on the volar surface of the forearm [28].

3.2 Alternately (simultaneous recording at different skin sites) recording SC and SP

Articles presented here are associated with authors, who tried to measure EDA parameters simultaneously, but not at the same skin site, i.e., EDA parameters were recorded alternately.

Wilcott [29] had recorded SR (SC) and SP simultaneously contralateral from the left and right-palm-upper-arm locations to study the correlation between both these EDA parameters. He obtained significant correlations between the SC and the SP amplitudes. Regarding the possibility of recording EDA parameters at the same skin site, Wilcott claimed that simultaneous measurement of EDA parameters is impossible and stated that "As it is of course not possible to record the two types of bioelectrical activity (e.g., SC and SP) from the same skin area simultaneously, they were recorded alternately from the same skin area and simultaneously from different skin areas" [29].

To elucidate mechanisms underlying the appearance of the diphasic potential (SP) curve, SC and SP simultaneously from the palm and the dorsum of the hand were recorded by Yokota et al. [30]. SC and SP simultaneously from the palm and the dorsum of the hand were recorded by Yokota et al. [30]. Sequence recording of SC and SP was applied so as to enable a comparison of the results with each other. When the SP was recorded simultaneously with SC from different parts of the same palm, diphasic potential curve changes usually corresponded to larger resistance changes [30].

According to Edelberg and Burch [31], both SC and SP are widely used as indicators of autonomic activity in psychological and pharmacological investigations, but the many fundamental controversies to be found in the literature testify to the questionable reliability of these measures [31]. In addition to the external stimuli and central excitatory state, various outside factors constitute a third category of variables (skin temperature, current density, electrode composition and size, contact medium, and electronic circuitry) affecting SP and SC responses. These factors are difficult to control, when comparisons between SP and SC are to be made separately (recording each of which at certain time). However, when they are compared simultaneously, effect of such variables can be canceled out [31].

Relations between simultaneously recorded SC and SP changes are obtained at high and low levels of bodily excitation by Darrow [22]. Electrodes were connected to the palm and the wrist of both hands in order to simultaneously record SC and SP between two opposite sites.

DOI: http://dx.doi.org/10.5772/intechopen.89025

Burstein et al. [32] reported simultaneous recording of SR (SC) and SP responses generated by different psychological stimuli. The aim was to show effects of such stimuli on responses of both EDA parameters and to determine the significance of the different wave forms of skin potential [32]. SC responses were recorded from middle and index fingers of the left hand, and SP responses were recorded from the right ear lobe and the right index finger. They found that both SC and SP are highly effective in assessing differential emotional responses. However, total SP yields slightly more significant results than does the SC measure. They finally suggested that the skin potential response merits further investigation because the qualitative differences in its wave forms may be of special significance [32].

Hupka and Levinger [33] recorded palmar SC and SP simultaneously with nonpolarizing electrodes, connected to the thenar eminence of the right and left palm and on the dorsal right or left forearm. Authors aimed to investigate whether the SPR negativity remains a constant correlation with the SCR during different conditions of passivity and motor activity [33].

SC and SP were recorded simultaneously from opposite hands during a stress period and a subsequent prolonged relaxation by Lykken et al. [34]. Their aim was to investigate certain interrelationships between concurrent measures of SC and SP through employing a simple equivalent circuit model of steady-state electrodermal phenomena [34].

The correlation of change measures and prestimulus level in SC and SP and the amount of correlation between SP and SC measures were studies by Gaviria et al. [35]. SP and SC were recorded in a sequence at two different skin sites.

The effect of repeated stimuli on reader and nonreader child with respect to physiological orienting response patterns in the autonomic nervous system was investigated by Hunter et al. [36]. Authors aimed to fined difference between those two groups through simultaneously recording SC and SP. For SC recording electrodes were placed to the volar pads of the first and third fingers, and for SP recordings the active electrode was placed on the volar pad of the middle finger, and the reference electrode was placed on the forearm; thus both SC and SP are recorded simultaneously at two different skin sites [36].

In order to investigate possible differences between healthy subjects and unhealthy (schizophrenic) subjects, Patterson and Venables [37] recorded SC and SP at the same time but at two different skin sites by two different measuring systems. Electrodes for SC recording were attached on the medial phalanges of the first and second fingers of both hands and for SP were placed on the hypothenar eminence of the right hand and an abraded surface of the right arm [37].

The validity of SC and SP for preliminary cystic fibrosis screening has been demonstrated by Williamson et al. [38]. Electrodes for SC recording were attached to the palmar surface of the fingers. For the measurements of SP, the active electrode was being placed on the palmar surface of the distal middle phalanx of the hand [38]. Discriminant analysis using the two best EDA measures (SC and SP) for assignment of experimental group membership yielded 92.7% correct classification of the actual group membership. However, both SP and SC recordings were necessary to obtain such accuracy, since reclassifications of participants by discriminant analysis using only the six SC or the six SP measures lowered the percentage of correct classifications to 77% for SC and 86% for SP. Authors stated that it would simplify clinical procedures to use only the SP or only the SC measures to distinguish between group (health and unhealthy) subjects, because only one hand would be recorded and scored [38]. However, according to their results, recording more than one EDA parameters at the same time would lead to more meaningful results.

Collet et al. [39] showed neutral and emotionally loaded pictures to some test subjects in order to evoke happiness, surprise, anger, fear, sadness, and disgust.

The EDA signals measured were SC, SP, and SR. For EDA recordings electrodes were placed at different skin sites. For SC measurement electrodes were attached to the second phalanx of the fourth and fifth digits of the non-dominant hand; for SP active electrode was attached to the hypothenar eminence, and the reference electrode was attached to the wrist, and finally for SR recording, electrodes were placed on the second phalanx of the index and the third digit of the non-dominant hand [39].

SR (SC) and SP and some other physiological parameters were recorded simultaneously at various skin sites by Ismaili et al. [40], to analyze the relationship between self-report hedonic evaluations and the physiological expression of emotion in response to odorants. Second phalanx of the index and the third digit of the non-dominant hand were selected for SC recording. For SP recording, the active electrode was placed on the hypothenar eminence, and the reference electrode was placed 10 cm higher on the wrist.

Shiihara et al. [41] examined the validity of long-term SC recordings by comparing such recordings with simultaneously recorded SP at two different skin sites. SC was recorded through placing electrode on the middle phalanx of the second and fourth fingers, and SP was recorded via attaching electrodes between the thenar eminence of the palm and the lower portion of the upper arm. Both EDA parameters were recorded simultaneously during sleep.

In a recent study, SC and SP have been recorded simultaneously at different skin sites to evaluate the mental workload during driving by Kajiwara [42]. Author stated that in order to simultaneously measure the SC and SP, the SC was measured on the left arm, and the SP was measured on the right arm.

3.3 Simultaneous measurement of SC and SP or SS at the same skin sites

As noted, earlier studies such as [29] have claimed that simultaneous measurements of EDA parameters are impossible. However, this hypothesis was later criticized and lost ground due to new explanations and newest trends and techniques.

Montagu in [43] performed measurements (SC and SP) on the same skin site simultaneously and continuously through the same pair of electrodes. Montagu aimed to study effects of variety of conditions (different external stimuli) on the simultaneous recordings of SC and SP at a single reacting area.

Grimnes [44] performed DC SP or SC, AC SC, and skin capacitance (i.e., SS) measurements on the same skin site simultaneously by using the three-electrode system. Author pointed out that the examination of the influence of sweat duct cannot be done without also recording the parallel values of the skin admittance. Also he reported that the sensitivity of the DC conductance to a certain reflex intensity is larger than that of the AC SC.

Qiao et al. [45] developed a method for simultaneously recording SC and SS at the same skin site via using a three-electrode lock-in amplifier measuring system. It is indicated that the use of AC excitation is necessary for recording SS.

The exogenous (SC) and endogenous (SP) responses in order to acquire a better understanding of the underlying mechanisms were compared by Jabbari et al. [46]. Both SCR and SPR were simultaneously recorded in the palms using the same electrodes. Authors reported independent information and correlation between SC and SP.

Grimnes et al. [47] developed a special recording system for combining SC with SP recordings from an active electrode at a palmar site together with a large indifferent electrode connected to a physiological NaCl bath in which the forearm was immersed. Their recording system used a small AC current, enabling the SC and SP to be recorded simultaneously at the same site. Authors reported that it is possible to measure the SP and skin AC SC simultaneously at the same skin site in a low-noise system. Authors pointed out that both the measuring systems that are used by Grimnes [44] and Montagu [43] had the demerit that the recording electrode must supply the necessary charge/discharge current to the blocking capacitor with changing DC voltages. This also introduced a time constant specified by the capacitance and the resistance of the recording electrode [47].

Pabst et al. [48] designed a measuring system for simultaneously recording AC SC and DC SC at the same skin site and same electrode. Authors aimed to examine the similarities and differences between the AC and DC methods of EDA recording under the same conditions. The measuring system consisted of two bipolar electrodes, which were placed at the thenar and hypothenar sites of the dominant hand.

3.4 Simultaneous measurement of SC, SP, and SS at the same skin sites

This type of recording must be done with AC exosomatic, since with DC exosomatic, SS could not be measured.

Jabbari et al. [9] developed a measuring system where DC current was replaced by a small AC current. They aimed to record SC, SP, and SS simultaneously at the same skin site. Three electrodes were placed to the skin, two measuring electrodes and one reference electrode [9]. They confirmed that it is possible to measure SP and skin AC SC simultaneously at the same skin site. In addition, they regarded recording SP and SC simultaneously with the same electrode as significant because of the often large skin site dependence of levels and response waveforms [9].

Tronstad et al. [8] used a new method to record SC, SP, and SS simultaneously at the same electrode. Their aim was to investigate the difference between waveform of SC and SP in a new way by comparing their temporal peak differences. It was a PC-based EDA recording system for simultaneous recording of skin admittance (SC and SS) and SP at the same electrode. For recoding EDA parameters, three electrodes were employed, a measuring electrode placed on hypothenar, a reference electrode placed on apex of elbow, and a current sink electrode placed on the underarm.

The effect of some external stimuli on simultaneous recordings of SC, SP, and SS at the same skin site has been investigated by Bari et al. [49]. Authors confirmed that the three EDA parameters (SC, SP, and S) could be measured simultaneously under the same electrodes. Each of such EDA parameters showed different wave forms depending on their relation with the tissue under investigation.

Figure 2 shows an example of a setup for simultaneously measuring SP, SC, and SS at the same skin site. It is a PC-based EDA system for recording EDA by means of skin admittance and SP simultaneously at the same electrode on the same skin site. It consists of one measuring electrode (ME), one reference electrode (RE), and a current sink electrode (CE).

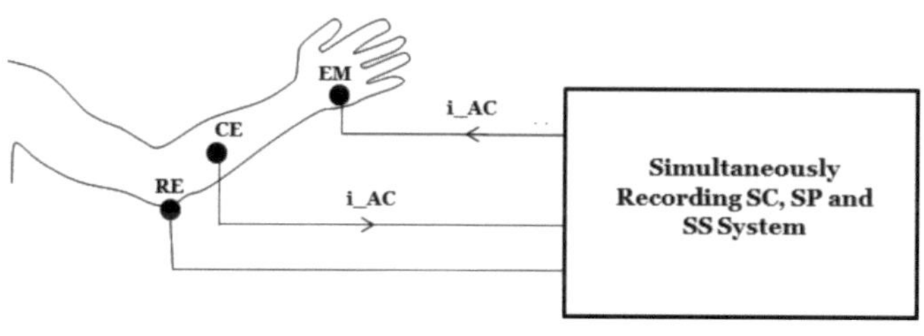

Figure 2.
An example of a setup for simultaneously recording SP, SC, and SS at the same skin site.

4. Advantages of simultaneous measurement of SC, SP, and SS over other EDA recording systems

1. In order to make direct comparisons between different EDA recording methods or parameters, techniques for simultaneously recording EDA using the same skin site are required, since EDA responses cannot be compared and/or correlated when lonely or separately measured.

2. Some authors such as [11] suggested a measuring method dependent on continuously switching (sequentially) between AC (SC) and DC (SP) recording methods, which allows the same site to be used for recording both SC and SP. However, this recording method has drawback that the switching skin to a different EDA coupler will always result in an adaptation of the gain process, dependent on the filter characteristics of the amplifier system, which constitutes a time-consuming procedure. In addition, the subject will seat or stay for longer time in order to record EDA parameters one by one, which lead to nuisance of the subject, and may in turn affect EDA signals (as EDA is associated with the psychophysiological state of the subject).

3. Also some other authors suggested another parallel-like (alternately) recordings of SC and SP measurements on contralateral sites [11]. But, this way of recording has disadvantage that the EDA parameters are not recorded at the same skin site. It has been pointed out that each active skin site may give rise to various types of EDA responses and that basic EDA levels all depend on the measuring system. So, measuring at the same electrode is significant because both EDA components (levels and response) are largely dependent on skin site as already mentioned above [47]. Besides the alternate method is inconvenient as both hands of the test subject are connected with the electrodes. In addition, to measure both SC and SP, at least four electrodes connected with two setups are required, which take economical cost. On the other hand, simultaneously employing only three electrodes that are placed on the same skin on a single hand makes it preferable and convenient where the subject will seat for shorter period than the other traditional recording methods. From an economic point of view, its cost is lower since one setup is used for recording all EDA parameters, and fewer electrodes are employed than other methods.

4. Simultaneous method also has another important advantage over the rest of the methods, which could be achieved when AC exosomatic current method is used. Through this method in addition to SC and SP, the SS could be recorded simultaneously at the same skin site as well. This means that EDA measurements can be simultaneously appropriate with additional psychophysiological measures, as it may provide further insights into the sweat gland physiology, which makes it particularly useful for clinical applications. Technically based upon this method, voltage sensing is converted from analog to digital, and utilizing a low measuring frequency with phase-sensitive rectification guarantees genuinely constant voltage or current and enables the minimization of measurement errors as well. Moreover, this recording method enables detection of variations in the reference site potential, thereby checking to which extent the reference site is electrodermally inactive, which is a requirement for accurate SP recording. Therefore, it is very suitable for physiological research.

5. Applications

The scope of applications will be primarily related to those studies recently conducted using the EDA measurement simultaneously. The use of simultaneous measurement of EDA in psychology, physiology, and medicine is widespread and constantly increasing due to its advantages. It is widely accepted that both AC and DC components (parameters) can be accurately measured.

Tronstad et al. [50] employed the system presented in Tronstad et al. [8], for the simultaneous recording of SC, SP, and SS at the same skin site. The aim of the study was to assess how accurately sweat production can be estimated based on combining the skin electrical properties. According to the authors, results of sweating estimation were significantly improved by the addition of SS and SP recordings to the SC recording only.

In a study [51], changes in SCR, SPR, and SSR were evaluated as a result of sequences of electrical (painful) stimuli with different intensities by using simultaneous system of EDA measurement. EDA responses as results of painful stimuli were recorded from 40 healthy volunteers. They reported that EDA responses significantly changed (increased) with respect to the intensity of the stimuli. Both SCR and SSR showed linear relationship with the painful stimuli. It was found that the EDA responses, particularly SCR ($p < 0.001$) and SSR ($p = 0.001$), were linearly affected by the intensity of the painful stimuli. Authors mentioned that EDA responses, in particular SCR, may be used as a useful indicator for assessment of experienced pain in clinical settings.

The same system mentioned above was used in another study [52], with the aim of exploring the influence of relative humidity on EDA levels and also the responses. A total of 10 healthy subjects were exposed to environments of low and high RH while EDA measures were recorded, including cognitive, visual, and breathing stimuli for evoking electrodermal responses of different origins. EDA levels and responses were compared between the two humidity levels for all stimuli and all EDA measures. It was found that EDA levels, in particular for SC and SS, were significantly increasing during high humidity exposure but that the change in EDA responses (SC, SS, and SP) was not statistically significant ($p > 0.05$, paired *t* test). Authors concluded that ambient humidity influences the recording of EDA levels and is important to consider when these parameters are used, but is not important in the recording or analysis of EDA responses.

6. Conclusions

The purpose of the chapter review was to view the trends in the methods and attempts made to develop them for recording EDA parameters. It was focused on the major methods (through some published literatures) used for recording EDA parameters. In addition, identifying the advantages of simultaneously recording EDA parameters at the same skin sites over sequential and alternate methods. Even though sequential or alternate recording methods certainly have advantages in some routine work, simultaneous recording method must be considered for future EDA research, since this measuring system can realize measurement of various EDA parameters on the human skin, at the same skin site and same time, which is required for accurate EDA research studies as well as clinical applications. This will in turn undoubtedly lead us to technological and clinical advancements in the treatment of other neurophysiological disorders. Therefore, simultaneous recording method is superior to sequential and alternate recording methods.

Conflict of interest

The authors declare no conflict of interest.

Acronyms and symbols

AC	alternating current
B	susceptance
C	electrical capacitance
CE	current sink electrode
DC	direct current
EDA	electrodermal activity
F	frequency
G	conductance
GSR	galvanic skin response
ME	measuring electrode
PC	personal computer
R	resistance
RE	reference electrode
RH	relative humidity
SC	skin conductance
SCL	skin conductance level
SCR	skin conductance response
SP	skin potential
SPL	skin potential level
SPR	skin potential response
SS	skin susceptance
SSL	skin susceptance level
SSR	skin susceptance response
SR	skin resistance
X	reactance
Y	electrical admittance
Z	electrical impedance

Author details

Haval Y. Yacoob Aldosky[1*] and Dindar S. Bari[2*]

1 Department of Physics, College of Science, University of Duhok, Duhok, Iraq

2 Department of Physics, Faculty of Science, University of Zakho, Zakho, Iraq

*Address all correspondence to: yacoobaldosky@uod.ac and dindar.bari@uoz.edu.krd

References

[1] Serup J, Jemec GB, Grove GL. Handbook of Non-Invasive Methods and the Skin. Boca Raton: CRC Press; 2006. DOI: 10.3109/9781420003307

[2] Humbert P, Fanian F, Maibach HI, Agache A. Measuring the Skin: Non-Invasive Investigation, Physiology, Normal Constants. 4th ed. Berlin: Springer; 2017. DOI: 10.1007/978-3-319-32383-1

[3] Dawson ME, Schell AM, Filion DL. The electrodermal system. In: Cacioppo JT, Tassinary LG, Berntson GG, editors. Handbook of Psychophysiology. Cambridge: Cambridge University Press; 2007. pp. 159-181. DOI: 10.1017/CBO9780511546396.007

[4] Edelberg R. Electrodermal mechanisms: A critique of the two-effector hypothesis and a proposed replacement. In: Roy JC, Boucsein W, Fowles DC, Gruzelier JH, editors. Progress in Electrodermal Research. NATO ASI Series (Series A: Life Sciences). Boston: Springer; 1993. pp. 7-29. DOI: 10.1007/978-1-4615-2864-7_2

[5] Edelberg R. Electrical activity of the skin: Its measurement and uses in psychophysiology. In: Greenfield NS, Stembach RA, editors. Handbook of Psychophysiology. New York: Holt, Rinehart & Winston; 1972. pp. 367-418

[6] Bera TK, Jampana N, Lubineau G. A LabVIEW-based electrical bioimpedance spectroscopic data interpreter (LEBISDI) for biological tissue impedance analysis and equivalent circuit modelling. Journal of Electrical Bioimpedance. 2016;7:35-54. DOI: 10.5617/jeb.2978

[7] Amini M, Hisdal J, Kalvøy H. Applications of bioimpedance measurement techniques in tissue engineering. Journal of Electrical Bioimpedance. 2018;**9**:142-158. DOI: 10.2478/joeb-2018-0019

[8] Tronstad C, Kalvøy H, Grimnes S, Martinsen ØG. Waveform difference between skin conductance and skin potential responses in relation to electrical and evaporative properties of skin. Psychophysiology. 2013;**50**:1070-1078. DOI: 10.1111/psyp.12092

[9] Jabbari A, Johnsen B, Grimnes S, Martinsen ØG. Simultaneous measurement of skin potential and conductance in electrodermal response monitoring. Journal of Physics: Conference Series. 2010;**224**:012091. DOI: 10.1088/1742-6596/224/1/012091

[10] Shira K, Yamamoto Y, Nakamura T, Kusuhara T. Formative mechanism of skin potential activity and relationships between skin potential and skin impedance. In: Proceedings of the IFMBE World Congress on Medical Physics and Biomedical Engineering; 2007. Berlin: Springer; 2007. pp. 2694-2697. DOI: 10.1007/978-3-540-36841-0_680

[11] Boucsein W. Electrodermal Activity. 2nd ed. New York: Springer Science & Business Media; 2012. DOI: 10.1007/978-1-4614-1126-0

[12] Tronstad C, Grimnes S, Martinsen ØG, Amundsen V, Wojniusz S. PC-based instrumentation for electrodermal activity measurement. Journal of Physics: Conference Series. 2010;**224**:012093. DOI: 10.1088/1742-6596/224/1/012093

[13] Fowles DC, Christie MJ, Edelberg R, Grings WW, Lykken DT, Venables PH. Publication recommendations for electrodermal measurements. Psychophysiology. 1981;**18**:232-239. DOI: 10.1111/1469-8986.ep11664429

[14] Greco A, Valenza G, Scilingo EP. Advances in Electrodermal Activity Processing with Applications for Mental Health: From Heuristic Methods to Convex Optimization. Cham: Springer; 2016. DOI: 10.1007/978-3-319-46705-4

[15] Grimnes S, Martinsen ØG. Bioimpedance and Bioelectricity Basics. 3rd ed. Oxford: Academic Press; 2014. DOI: 10.1016/C2012-0-06951-7

[16] Benedek M, Kaernbach C. Decomposition of skin conductance data by means of nonnegative deconvolution. Psychophysiology. 2010;**47**:647-658. DOI: 10.1111/j.1469-8986.2009.00972.x

[17] Greco A, Lanata A, Valenza G, Rota G, Vanello N, Scilingo EP. On the deconvolution analysis of electrodermal activity in bipolar patients. In: Proceedings of the IEEE Annual International Conference on Engineering in Medicine and Biology Society (Embc); 2012. San Diego: IEEE; 2012. pp. 6691-6694. DOI: 10.1109/EMBC.2012.6347529

[18] Greco A, Valenza G, Lanata A, Rota G, Scilingo EP. Electrodermal activity in bipolar patients during affective elicitation. IEEE Journal of Biomedical and Health Informatics. 2014;**18**:1865-1873. DOI: 10.1109/JBHI.2014.2300940

[19] Venables PH, Christie MJ. Electrodermal activity. In: Martin I, Venables PH, editors. Techniques in Psychophysiology. Chichester: Wiley; 1980. pp. 3-67. DOI: 10.2307/1422756

[20] Neumann E, Blanton R. The early history of electrodermal research. Psychophysiology. 1970;**6**(4):453-475. DOI: 10.1111/j.1469-8986.1970.tb01755.x

[21] Tronstad C, Gjein GE, Grimnes S, Martinsen ØG, Krogstad AL, Fosse E. Electrical measurement of sweat activity. Physiological Measurement. 2008;**29**:S407-S415. DOI: 10.1088/0967-3334/29/6/S34

[22] Darrow CW. The rationale for treating the change in galvanic skin response as a change in conductance. Psychophysiology. 1964;**1**:31-38. DOI: 10.1111/j.1469-8986.1964.tb02618.x

[23] Lykken DT, Venables PH. Direct measurement of skin conductance: A proposal for standardization. Psychophysiology. 1971;**8**:656-672. DOI: 10.1111/j.1469-8986.1971.tb00501.x

[24] Edelberg R. Electrical properties of skin. In: Elden HR, editor. A Treatise of the Skin (Biophysical Properties of the Skin). New York: Wiley; 1971. pp. 519-551

[25] Fowles DC. The eccrine system and electrodermal activity. In: Coles MGH, Donchin E, Porges SW, editors. Psychophysiology: Systems, Processes, and Applications. Amsterdam: Elsevier; 1986. pp. 51-96

[26] Venables P, Sayer E. On the measurement of the level of skin potential. British Journal of Psychology. 1963;**54**:251-260. DOI: 10.1111/j.2044-8295.1963.tb00880.x

[27] Venables P, Martin I. The relation of palmar sweat gland activity to level of skin potential and conductance. Psychophysiology. 1967;**3**:302-311. DOI: 10.1111/j.1469-8986.1967.tb02710.x

[28] Turpin G, Siddle DA. Effects of stimulus intensity on electrodermal activity. Psychophysiology. 1979;**16**:582-591. DOI: 10.1111/j.1469-8986.1979.tb01525.x

[29] Wilcott R. Correlation of skin resistance and potential. Journal of Comparative and Physiological Psychology. 1958;**51**:691. DOI: 10.1037/h0038675

[30] Yokota T, Takahashi T, Kondo M, Fujimori B. Studies on the diphasic wave form of the galvanic skin reflex. Electroencephalography and Clinical Neurophysiology. 1959;**11**:687-696. DOI: 10.1016/0013-4694(59)90109-9

[31] Edelberg R, Burch NR. Skin resistance and galvanic skin response: Influence of surface variables, and methodological implications. Archives of General Psychiatry. 1962;7:163-169. DOI: 10.1001/archpsyc.1962.01720030009002

[32] Burstein KR, Fenz WD, Bergeron J, Epstein S. A comparison of skin potential and skin resistance responses as measures of emotional responsivity. Psychophysiology. 1965;**2**:14-24. DOI: 10.1111/j.1469-8986.1965.tb02630.x

[33] Hupka RB, Levinger G. Within-subject correspondence between skin conductance and skin potential under conditions oe activity and passivity. Psychophysiology. 1967;**4**:161-167. DOI: 10.1111/j.1469-8986.1967.tb02754.x

[34] Lykken D, Miller R, Strahan R. Some properties of skin conductance and potential. Psychophysiology. 1968;**5**:253-253-268. DOI: 10.1111/j.1469-8986.1968.tb02821.x

[35] Gaviria B, Coyne L, Thetford PE. Correlation of skin potential and skin resistance measures. Psychophysiology. 1969;**5**:465-477. DOI: 10.1111/j.1469-8986.1968.tb02821.x

[36] Hunter EJ, Johnson LC, Keefe FB. Electrodermal and cardiovascular responses in nonreaders. Journal of Learning Disabilities. 1972;**5**:187-197. DOI: 10.1177/002221947200500402

[37] Patterson T, Venables PH. Bilateral skin conductance and skin potential in schizophrenic and normal subjects: The identification of the fast habituator group of schizophrenics. Psychophysiology. 1978;**15**:556-560. DOI: 10.1111/j.1469-8986.1978.tb03109.x

[38] Williamson PS, Fowles DC, Weinberger M. Electrodermal potential and conductance measurements clinically discriminate between cystic fibrosis and control patients. Pediatric Research. 1985;**19**:810-814. DOI: 10.1203/00006450-198508000-00006

[39] Collet C, Vernet-Maury E, Delhomme G, Dittmar A. Autonomic nervous system response patterns specificity to basic emotions. Journal of the Autonomic Nervous System. 1997;**62**:45-57. DOI: 10.1016/S0165-1838(96)00108-7

[40] Alaoui-Ismaïli O, Vernet-Maury E, Dittmar A, Delhomme G, Chanel J. Odor hedonics: Connection with emotional response estimated by autonomic parameters. Chemical Senses. 1997;**22**:237-248. DOI: 10.1093/chemse/22.3.237

[41] Shiihara Y, Umezawa A, Sakai Y, Kamitamari N, Kodama M. Continuous recordings of skin conductance change during sleep. Psychiatry and Clinical Neurosciences. 2000;**54**:268-269. DOI: 10.1046/j.1440-1819.2000.00672.x

[42] Kajiwara S. Evaluation of driver's mental workload by facial temperature and electrodermal activity under simulated driving conditions. International Journal of Automotive Technology. 2014;**15**:65-70. DOI: 10.1007/s12239-014-0007-9

[43] Montagu J. The psycho-galvanic reflex; a comparison of A.C. skin resistance and skin potential changes. Journal of Neurology, Neurosurgery, and Psychiatry. 1958;**21**:119-128. DOI: 10.1136/jnnp.21.2.119

[44] Grimnes S. Psychogalvanic reflex and changes in electrical parameters of dry skin. Medical &

Biological Engineering & Computing. 1982;**20**:734-740. DOI: 10.1007/BF02442528

[45] Qiao Z-G, Mørkrid L, Grimnes S. Simultaneous measurement of electrical admittance, blood flow and temperature at the same skin site with a specially designed probe. Medical & Biological Engineering & Computing. 1987;**25**:299-304. DOI: 10.1007/BF02447428

[46] Jabbari Z, Grimnes S, Martinsen ØG. Electrodermal response–correlation between potential and conductance. In: Proceedings of the IFMBE 13th International Conference on Electrical Bioimpedance and the 8th Conference on Electrical Impedance Tomography; 2007. Berlin: Springer; 2007. pp. 747-750. DOI: 10.1007/978-3-540-73841-1_193

[47] Grimnes S, Jabbari A, Martinsen ØG, Tronstad C. Electrodermal activity by DC potential and AC conductance measured simultaneously at the same skin site. Skin Research and Technology. 2011;**17**:26-34. DOI: 10.1111/j.1600-0846.2010.00459.x

[48] Pabst O, Tronstad C, Grimnes S, Fowles D, Martinsen ØG. Comparison between the AC and DC measurement of electrodermal activity. Psychophysiology. 2016;**54**:374-385. DOI: 10.1111/psyp

[49] Bari DS, Yacoob HY, Tronstad C, Kalvøy H, Martinsen ØG. Electrodermal responses to discrete stimuli measured by skin conductance, skin potential and skin susceptance electrodermal responses. Skin Research and Technology. 2018;**24**:108-116. DOI: 10.1111/srt

[50] Tronstad C, Kalvøy H, Grimnes S, Martinsen ØG. Improved estimation of sweating based on electrical properties of skin. Annals of Biomedical Engineering. 2013;**41**:1074-1083. DOI: 10.1007/s10439-013-0743-4

[51] Bari DS, Yacoob HY, Tronstad C, Kalvøy H, Martinsen ØG. Electrodermal activity responses for quantitative assessment of felt pain. Journal of Electrical Bioimpedance. 2018;**9**:52-58. DOI: 10.2478/joeb-2018-0010

[52] Bari DS, Yacoob HY, Tronstad C, Kalvøy H, Martinsen ØG. Influence of relative humidity on electrodermal levels and responses. Skin Pharmacology and Physiology. 2018;**31**:298-307. DOI: 10.1159/000492275

Chapter 7

Experimental and Theoretical Study of the Adsorption Behavior of Nitrate Ions by Layered Double Hydroxide Using Impedance Spectroscopy

Abderrahmane Elmelouky, Abdelhadi Mortadi, Elghaouti Chahid and Reddad Elmoznine

Abstract

This chapter analyzes the experimental data using impedance spectroscopy to reduce water pollution by nitrate ions. The adsorption is through a synthesized layered double hydroxide (Zn_3-Al-Cl-LDH). The kinetic study data analysis by pseudo-first-order and pseudo-second-order models is highly correlated they were found to fit very well the pseudo-second-order. This is confirmed by fast kinetic modeling of experimental data according to the pseudo-second-order. Furthermore, the Nyquist plots suggest that the grains and grain boundaries have contributed to the conduction mechanism of the material at different adsorption times and monitoring of the adsorption phenomenon. The investigation by impedance spectroscopy was used for modeling by an equivalent circuit. The real and imaginary functions of impedance complex are analyzed by modifying Cole-Cole relaxation. Revel most changes in the structure of the manifestation of the grains and the grains boundaries. The alternative current (AC) conductivity was investigated using the double power law of Jonscher. More importantly, the calculated value and the percentage of efficiency are evaluated in the adsorption. The water molecules and nitrate ions in the adsorbed were favored for the generation of the electrical response. The electrochemical impedance spectroscopy data are often interpreted by using electrical equivalent circuits.

Keywords: materials science, materials chemistry, impedance spectroscopy, layer charge, ionic, conductivity, dielectric

1. Introduction

Impedance spectroscopy is used to obtain an electrical equivalent circuit that showed the best fit of the experimental data. The analysis of these impedance data using both Bode and Nyquist plots was useful for the identification and the deconvolution of different processes of adsorption and interfaces. The adsorption-desorption phenomena representing the fundamental and essential electrochemical processes at the solid-liquid interface and adsorption of ionic and nonionic surfactants on the solid/liquid interfaces have been studied experimentally [1–4] and theoretically [5–8]. The surface area of a solid electrode also affects adsorption efficacy [3]. On the other hand, the contamination of wastewater by nitrate ions has become an ever-increasing and serious environmental threat for a long time [9]. The disproportionate application of chemical products in industrial sectors causes the increases in large quantities of this ion into wastewater and surface water [10]. The solubility of the ions of nitrate is very high in water [11] is considered the most widespread contaminant exhibit a serious in this literature [12, 13]. The high amount of nitrate in drinking water can cause a number of health problems such as gastric cancer [14]. The methemoglobinemia or blue baby syndrome, a serious health danger, occurs when nitrate is converted to nitrite which then reacts with the hemoglobin to cause blueness of the skin of newborn infants; this is discussed by [15]. The superior costs of adsorption using adsorbents such as activated carbon prompt researchers to find other cheaper substitutes such as lamellar double hydroxides (LDH) which have been shown to be effective for the removal of this contaminant from industrial washing water [16–18]. Although studies have examined this material, its application toward the removal of nitrate has not been examined previously. Therefore, the present chapter is aimed to synthesize and to study the efficient removal of nitrate by LDH. This system has low-mobility charge carriers, which are believed to be responsible for the dielectric response.

2. Equipment and its use

2.1 Synthetic adsorbent and characterization techniques

The coprecipitation method was used to prepare Zn_3-Al-Cl-LDH described by [19], with constant pH (pH = 7) and molar ratio (R = Zn^{2+}/Al^{3+}). Pure $ZnCl_2$ (0.3 mol) and $AlCl_3.6H_2O$ (0.1 mol) solution with a molar ratio of R = 3 is dissociated in distilled water (250 mL) to produce solution (a) and solution (b) containing NaOH (0.8 mol) and NaCl (0.05 mol) in 250 mL of distilled water. After solution (b) is added to solution (a) with vigorous stirring at pH = 7 with an N_2 start to control contamination by carbonate ions. The precipitate was thoroughly washed with deionized water and dried at 60°C for 24 h to obtain the LDH of formula ($Zn_{2.93}$Al (OH) 7.86) (Cl^-. 1.87 H_2O noted Zn_3Al-Cl-LDH according to [20].

In the X-ray powder diffraction (XRD), the samples were recorded on an X-ray diffractometer (SIEMENS D 501) with a radiation of $\lambda K\alpha 1$ = 1.5405 Å and $\lambda K\alpha 2$ = 1.5444 Å. The patterns were recorded from 2° to 76° 2θ angles at a step size of 0.02° and at a speed of 5°/min.

The Perkin-Elmer 16 PC Fourier transform spectrometer (FTS) was used for infrared measurements. The samples were prepared in a pellet of 13 mm diameter and 1 mm thickness using 2 mg of product diluted in 200 mg of KBr. The FT-IR spectra were recorded in absorbance in the wave number range of 400–4000 cm^{-1} at 25°C with a resolution of 1 cm^{-1}.

The frequency range from 20 Hz to 1 MHz was performed for impedance spectroscopy measurements, with eight points per decade at room temperature utilizing an impedance analyzer (Hewlett Packard 4192A). The electrical contacts were performed using silver electrodes, which were deposited on the two circular faces of the sample [19]. The magnitude of the applied signal is 0.6 V peak to peak. An amount of 200 mg is pelleted to analyze the impedance [19]. The granulated powder was compacted under a hydraulic press with 250 MPa pressure into discs of 13 mm diameter and 1 mm thickness approximately [19]. The impedance spectra were recorded at different adsorption times (5, 10, 20, 30, and 60 min). The analysis and theoretical fitting by impedance spectroscopy using complex empirical functions were carried out utilizing the software ZView 2.2 and Origin 8 for modeling of the ionic conductivity and the imaginary function according to the real function, respectively.

3. Results and discussion

3.1 X-ray diffraction

The X-ray diffraction patterns of Zn_3-Al-Cl-LDH depicted in **Figure 1** of the sample are characteristic to those of a double lamellar hydroxide. The sample was crystallized in a rhombohedral symmetry (space group: R-3m) with (c/3) = d_{003} = $2d_{110}$ and a = $2d_{006}$. The lattice parameters **c** and **a** are, respectively, 2.38 and 0.31 nm. These values are similar to those reported in the literature [14].

The peak (1 1 0) indicates the intermetallic distance used to calculate a lattice parameter (a = 2d110). Moreover, the values of the parameters **c** and **a** are, respectively, 23.82 and 3.10 A°. These values are similar to those reported in the literature [19].

3.2 Infrared spectroscopy

FT-IR confirms that the spectra of synthesized Zn_3Al-Cl-$(NO_3^-)_{ads}$-LDH (**Figure 2**) resemble those of hydrotalcite-like phases [21]. The FT-IR presents a close-up view of the most important regions of the infrared spectra of Zn_3Al-Cl-$(NO_3^-)_{ads}$-LDH depicted in **Table 1**.

The frequencies of absorbance links in this material are reported in **Table 1**. Indeed the infrared spectra of this material after adsorption at a different time show the increase in the intensity of characteristic link of NO_3^- ions at 1381 cm^{-1} as a function of time.

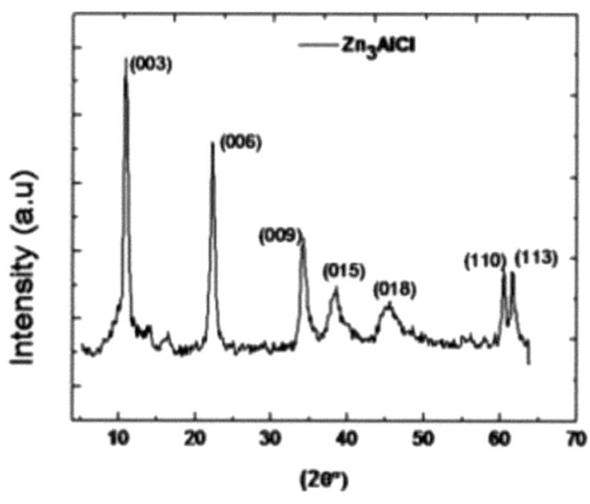

Figure 1.
XRD pattern of Zn_3-Al-Cl-LDH-blank.

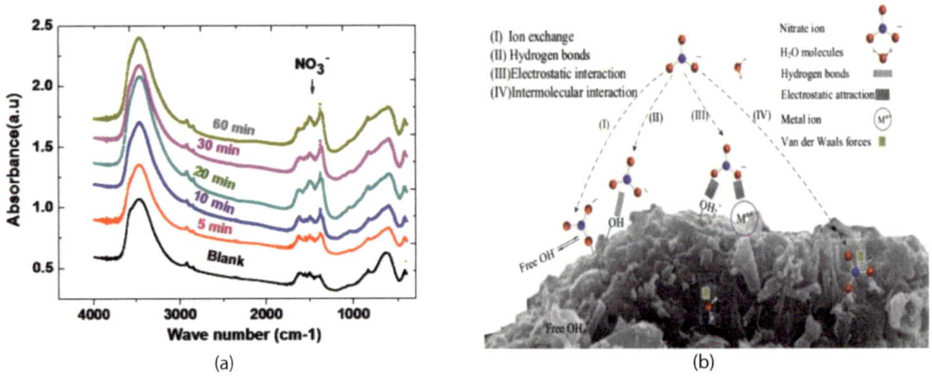

Figure 2.
(a) FT-IR spectra of Zn_3-Al-Cl visualized and (b) proposed adsorption mechanism between the adsorbent and nitrate [22].

Sample	Ratio	υ (OH) (cm^{-1})	δ (H_2)) (cm^{-1})	ν (M O) (cm^{-1})	δ (O M O) (cm^{-1})
Zn_3-Al-Cl-LDH-blank	R = 3	3454.6	1628	613.38	426.28

Table 1.
Frequencies of absorbance bands.

3.3 Inductively coupled plasma (ICP) spectrometry

The chemical formula Zn_3-Al-Cl-LDH was obtained using the technical ICP analysis, which shows that the theoretical ratio [R = (Zn^{2+}/Al^{3+})] is close to that of the synthesis. This characterization also suggests that the sample has a homogeneous chemical composition; the approximate chemical formula is ($Zn_{2.93}$Al $(OH)_{7.86}$) (Cl^-. 1.87 H_2O) for the metal ratio of R = 3. Cl^- anion intercalated.

3.4 Study of the adsorption in batch and determination of nitrate ions

3.4.1 Calibration curve

A series of vials of 50 ml each containing 1 mL of 0.5% sodium salicylate is introduced: 8, 6, 4, and 2 ml of KNO_3 solution (50 mg L^{-1}) with, respectively, 2, 4, 6, and 8 ml of water. The contents of each flask are then evaporated to dryness in an oven (75–80°C) to produce a residue which is dissolved with 2 ml of concentrated H_2SO_4. After 10 min of rest has passed, 15 ml of distilled water and then 15 ml of a basic solution made up of 40% NaOH, 6% sodium potassium tartrate, and then 50 ml with distilled water are then allowed to develop the yellow color characteristic of the complex nitrate formed. The nitrate ion concentrations in the liquid phase were determined by a spectrophotometric method (spectrophotometer type JASCOV-630 λ = 415 nm).

The calibration curve obtained (**Figure 3**) is in good agreement with the Beer-Lambert law.

3.4.2 Assays of nitrate ions in balance

The Zn_3Al-Cl-LDH is contacted with a potassium nitrate solution (0.4 g/L) for 5 min, 10 min, 20 min, 30 min, and 60 min. Before washing Zn_3Al-Cl-LDH after

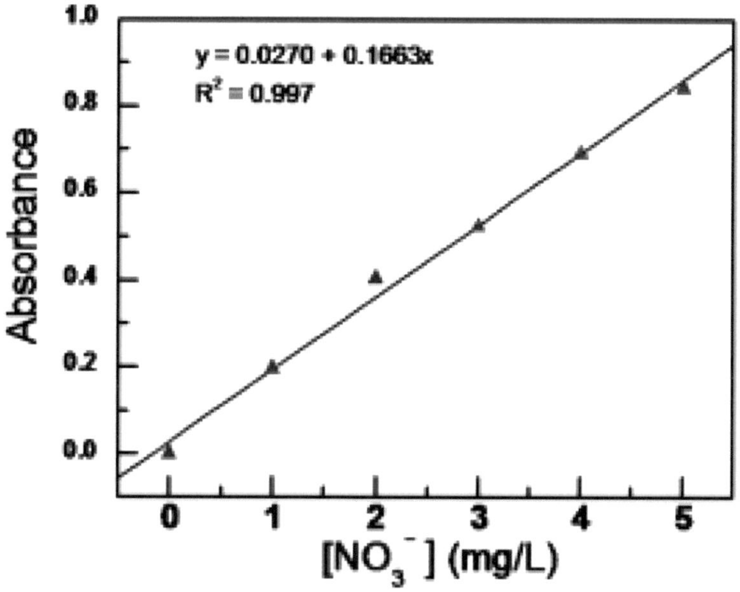

Figure 3.
Curve calibration of spectrophotometer used for determination of NO_3^- adsorbed.

adsorption and which is intended for the dielectric analysis. The filtrate is recovered corresponding to each time and fee in 1 mL is put in a 50 mL flask and then the following procedure used for the tracing of the calibration curve.

3.5 Kinetic study

The adsorption kinetic studies were carried out by contacting Zn_3Al-Cl-LDH (C_m = 0.8 g/L) with NO_3^- solutions (500 ml) of the initial concentration of 0.4 mg/L, respectively. The adsorption process was agitated at 25°C and a pH of 7.0 for several periods ranging from 5 to 60 min under inert atmosphere (N_2). LDH obtained after adsorption was filtered and then washed several times. The concentration of the nitrate ion in the filtrate was determined by spectrophotometer at 415 nm. The nitrate amount q_e (mg/g) loaded on adsorbents after adsorption experiments and the percentage removal (removal %) of NO_3^- ions from solutions were calculated using the following equations:

$$q_e = \frac{(C - Ce) \times V}{m} \tag{1}$$

and the removal

$$(\%) = \frac{C_0 - C_e}{C_0} \times 100 \tag{2}$$

where Ce (mg/L) is the equilibrium of nitrate ion concentration in solution, C_0 (mg/L) is the initial of nitrate ion concentration in solution, m (g) is the mass of adsorbent, and V (L) is the volume of the solutions.

The equilibrium is reached after 30 min, with a maximum of approximately 59.12% adsorption capacity corresponding to a 295.62 mg/g of an affinity of the adsorbate for the active sites of the adsorbent [23]. From **Figure 3** it is quite clear that the percentage of nitrate ion adsorption calculated by kinetic study and

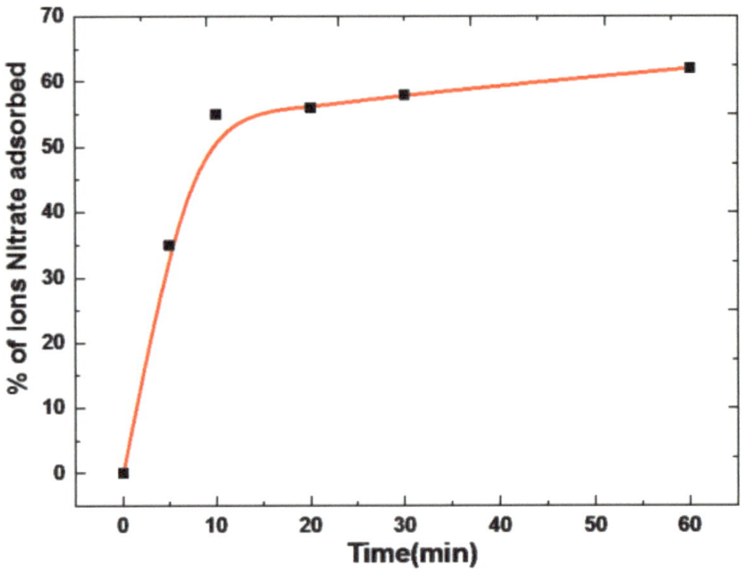

Figure 4.
Kinetic study evaluation of NO_3^- removal by Zn_3AlCl-$(NO_3^-)_{ads}$ complex.

efficiency adsorption calculated by impedance spectroscopy are less than 10% which shows that the values are close and both techniques are okay and best correlated. From **Figure 4** adsorption of nitrate ions by this system noted as a function of adsorption time is quite rich on information. The adsorption phenomena are due to the active sites in LDH interlayer with different electron donor sites (active adsorption sites) on the ions (N&O) and relative humidity [24–26].

3.6 Analysis of adsorption kinetics

The different kinetic models including pseudo-first-order, pseudo-second-order, and intraparticle diffusion are employed to investigate the mechanism of adsorption and potential rate controlling steps such as chemical reaction mass transport and diffusion control processes [27]. The pseudo-first-order and pseudo-second-order are generally expressed as Eqs. (3) and (4), respectively (**Figure 5**):

$$\text{Log}\left(q_e - q_t\right) = \text{Log}\left(q_e\right) - \frac{K_{1,ads}}{2,303} \times t \tag{3}$$

$$\frac{t}{q_t} = \frac{1}{K_{2,ads} \times q_e^2} + \frac{t}{q_e} \tag{4}$$

where qe (mg/g) and q_t (mg/g) are the adsorption of NO_3^- ions on to adsorbents at equilibrium and at time t (min), respectively; $K_{1,ads}$ (min^{-1}) and $K_{2,ads}$ (g/(mg min)) are the constants of the pseudo-first-order and pseudo-second-order adsorption, respectively. Additionally h (mg/(g min)) is the initial adsorption rate of pseudo-second-order which can be calculated using h = K_2qe2.

The adsorption rate constants $k_{1,ads}$ and $k_{2,ads}$ of nitrate ions by (Zn_3Al-Cl-LDH) are deducted, respectively, from the curve log(q_e-q) = f(t) and (t/q_t) = f(t/q_e) (**Figures 6** and 7).

The regression by the pseudo-second-order model agrees well to study the adsorption of the nitrate ions by Zn_3Al-Cl-LDH. The constant of adsorption rate

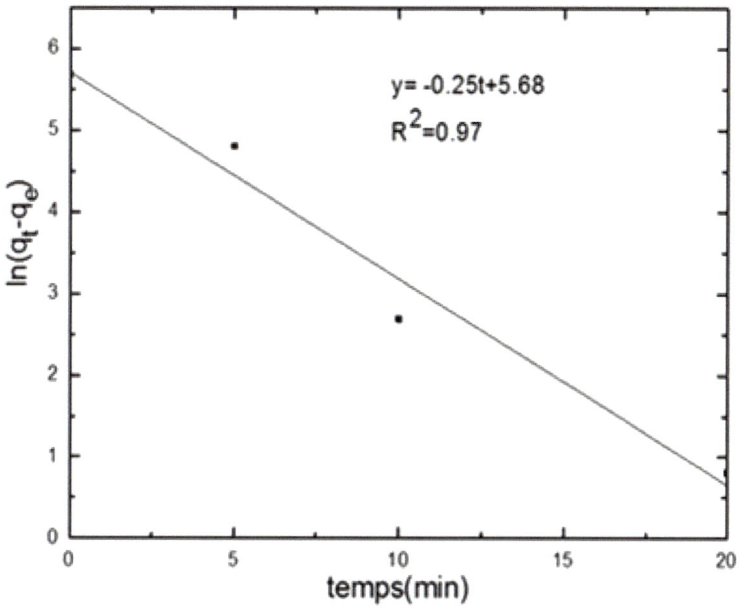

Figure 5.
Pseudo-first-order model removal NO_3^- by (Zn_3Al-Cl-LDH).

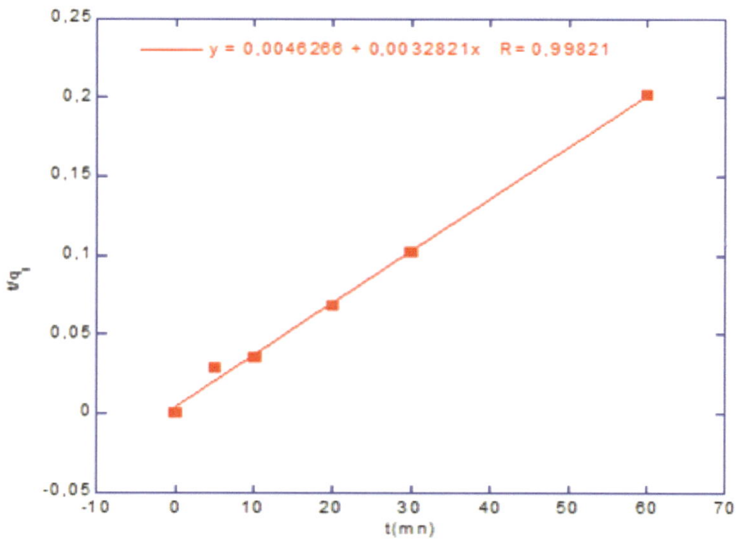

Figure 6.
Pseudo-second-order model of removal NO_3^- by Zn_3-Al-Cl-LDH.

confirms the rapid process noted during the kinetic study. The maximum amount obtained by applying the pseudo-second-order model is very close to that determined by the kinetic study (~296 mg/g).

The values of the correlation factors obtained (**Table 2**) show that the measurement data of kinetics follows the pseudo-second-order model ($R^2 \approx 0.99$).

3.7 Constant diffusion rate determination

Intraparticle diffusion equation suggests that intraparticle diffusion is the rate-limiting step in adsorption. The diffusion process may affect the adsorption of

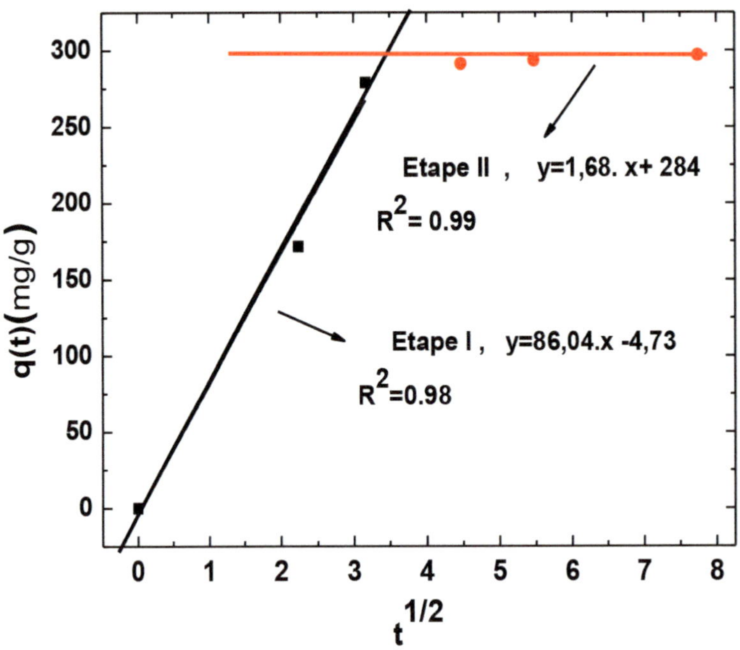

Figure 7.
Intraparticle diffusion model for $Zn_3Al\text{-}Cl\text{-}(NO_3^-)_{ads}$-LDH complex.

Pseudo-first-order			Pseudo-second-order		
$K_{1.ads}$ (min^{-1})	qe (mg/g)	R^2	$K_{2.ads}$ (g/mg/min)	qe (mg/g)	R^2
0.25	292.94	0.94	0.002	312.5	0.99

Table 2.
Kinetic parameters for $Zn_3Al\text{-}Cl\text{-}(NO_3^-)_{ads}$-LDH complex.

nitrate ions on Zn_3-Al-Cl-LDH due to the porous structure of the adsorbent and the attractive effect of nitrate ions. Therefore the intraparticle diffusion is used to explore the behavior of intraparticle diffusion is obeys Eq. (5) [27].

$$q_t = kip.t^{1/2} + C \quad (5)$$

where qt is the quantity retained at time t and kip are the diffusion rate constants.

The results obtained (**Figure 6**) show that there are two stages. *Region 1* is attributed to the most readily available site on the surface of the adsorbent. *Region 2* can be explained by a very slow diffusion of adsorption in the inner pores. Thus the nitrate ion adsorption by Zn_3-Al-Cl-LDH may be governed by the intraparticle model [1]. The values of kp1 and kp2 diffusion rate constants for Region 1 and Region 2, respectively, obtained by using the regression linear are shown in **Table 2**.

These values are in good agreement with the kinetic study. Indeed, in Region 1 the value of the slope (86.04) is greater than that of the Region 2 whose value is of the order of 1.68. This can be explained by the availability of sites in Zn_3-Al-Cl-LDH at the beginning of adsorption. The release rate constants intraparticle using the kinetics study was according the values respectively K_{1P} = 86.04 with the R^2 = 0.98 and K_{2P} = 1.68 with the R^2 = 0.99. The adsorption of nitrate ions by Zn_3-Al-Cl-LDH is confirmed by FT-IR spectroscopy. In fact the infrared spectra of the materials recovered after adsorption at various times (**Figure 6**) show the increase of the intensity of the characteristic band of NO_3^- at 1381 cm^{-1} in a function of contact time of ions.

It could be seen that the plots were multilinear over the whole time range suggesting that two steps were operational in the adsorption of NO_3^- by Zn_3-Al-Cl-LDH. The first linear plot was the instantaneous adsorption or external surface adsorption attributing to the rapid consumption of the available adsorption sites on the adsorbent surface. The second stage was the gradual adsorption stage where the intraparticle was the rate-limiting step, and the second portion was attributed to the final equilibrium for which the intraparticle diffusion starts to slow down due to the extremely few adsorption sites left on adsorbent which will be clearly in impedance spectroscopy using the Nyquist diagram analysis by means of fit and extrapolation of experimental data for both adsorption regions.

4. Impedance spectroscopy analysis

Using impedance spectroscopy one can distinguish between intrinsic (grain) and extrinsic (grain boundaries). The Nyquist and Bode plots were used to interpret the electrical relaxation processes associated with adsorption phenomenon and used to search for the electrical analogue of the adsorption.

In order to determine the mechanisms responsible for the conductivity, in general, the different variations of complex impedance spectra (Nyquist plot) consist of two semicircle arcs corresponding to the grain interior and grain boundary. The arc at a high frequency usually represents the grain response, and the low-frequency arc corresponds to the grain boundary [28].

The impedance analysis allows one to determine the contributions of various processes such as bulk effects and the grain boundaries. **Figure 8** shows the complex impedance plane plots (Nyquist plot) of the nitrate ion removal by the system Zn_3-Al-Cl-LDH (0 min $\leq$ t $\leq$ 60 min).

The analysis (**Figure 8a** and **b**) of the data by the Nyquist diagram allowed us to determine the resistance values for the two regions in the time interval of 0, 5, and 10 min (Region 1) and 20, 30, and 60 min (Region 2). The separation into two time intervals is justified by **Figure 7** of the kinetic study and **Figures 9** and **10** of the impedance spectroscopy study.

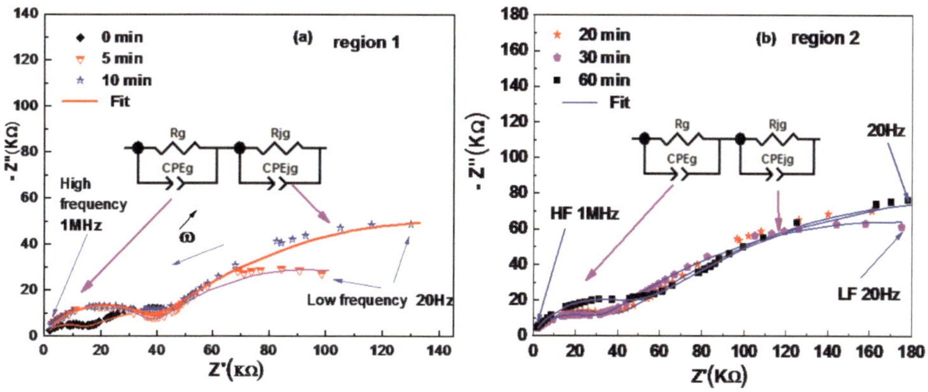

Figure 8.
Nyquist plots for samples in Region 1 (a) and Region 2 (b), respectively, during adsorption phenomenon by Zn_3Al-Cl-$(NO_3^-)_{ads}$-LDH.

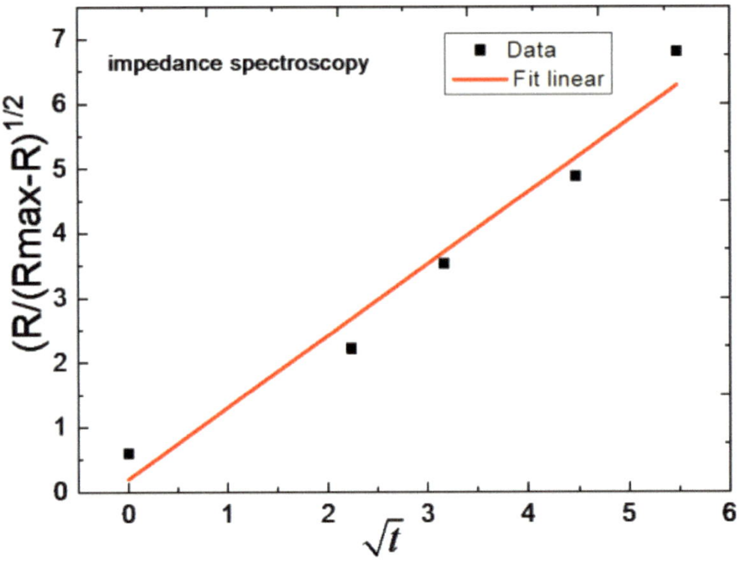

Figure 9.
Variation of the ratio (R_{bg}/R_{bgmax}-R_{bg}) according to the square root of the adsorption time.

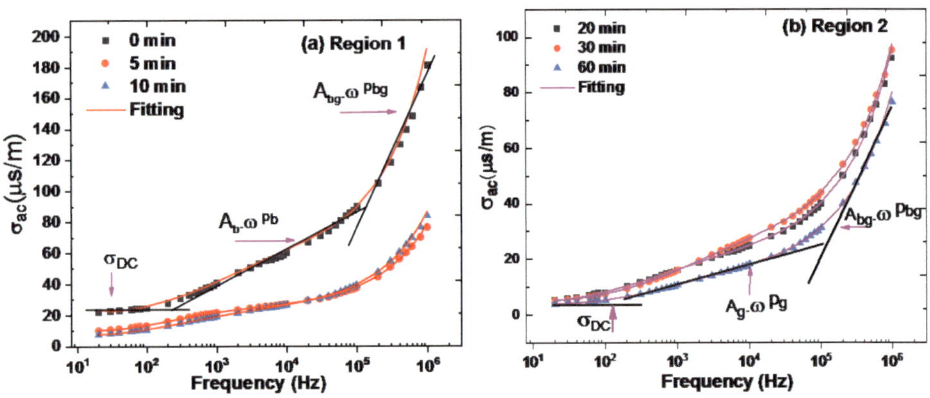

Figure 10.
Variation of σ'ac conductivity for (a) Region 1 and (b) Region 2 as a function of frequency.

Figure 11 shown is intended to show that there are only two capacitive loops in the extrapolation technique of experimental data in a frequency range of 10^{-2} Hz to 1 MHz: one for the contribution of grain region and the other for grain boundary contribution.

The estimated value of R is the difference between the high intercept and low intercept values. If the data reflects a parallel R-C element with a depression angle of zero, the estimated R will be the same as the diameter of the semicircle. Estimated C is calculated using the relationship ω_{max} = 1/(RC) and depends on the accuracy of ω_{max}.

The Nyquist (**Figures 12** and **13**) plot studies discovered the presence of grain and grain boundary which then become dipoles when they are subjected to the action of an electric field. We thus observe two phenomena of relaxations, no more of which are observed as the maximum in the Nyquist plot (Cole-Cole). We also find that we have two different regions that are two constants of the time τ_g and τ_{bg} using extrapolation by a corresponding equivalent circuit of data, which leads us to say this kinetic is mixed and it is what we confirmed by the chemical kinetic study.

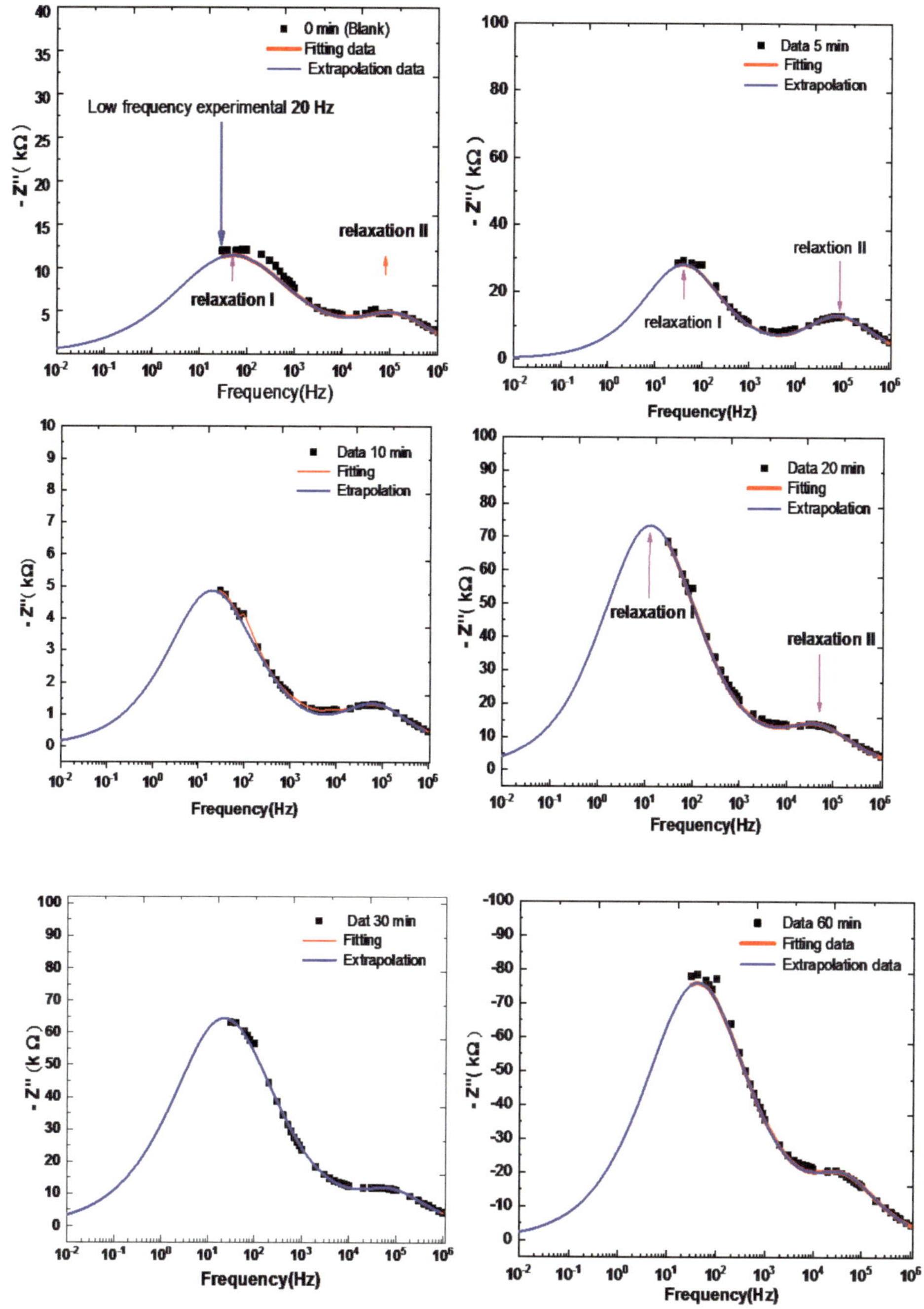

Figure 11.
Superposition of experimental data done by the equivalent electrical circuit and extrapolation of different adsorption times.

After analyzing and evaluating the spectra using ZView 2.2 software, we extracted the parameters mentioned in **Table 3**. From the values shown in **Table 3**, it is observed that the adsorption efficiency increases when showing the fixing of nitrate ions on the surface of Zn_3-Al-Cl-LDH. We used other quantities extracted from the equivalent circuit. The quantities are the dispersion coefficient for the grain and the grain boundary, on the one hand, and the grain and grain seal capacity, on the other hand, as a function of the adsorption time in order to follow

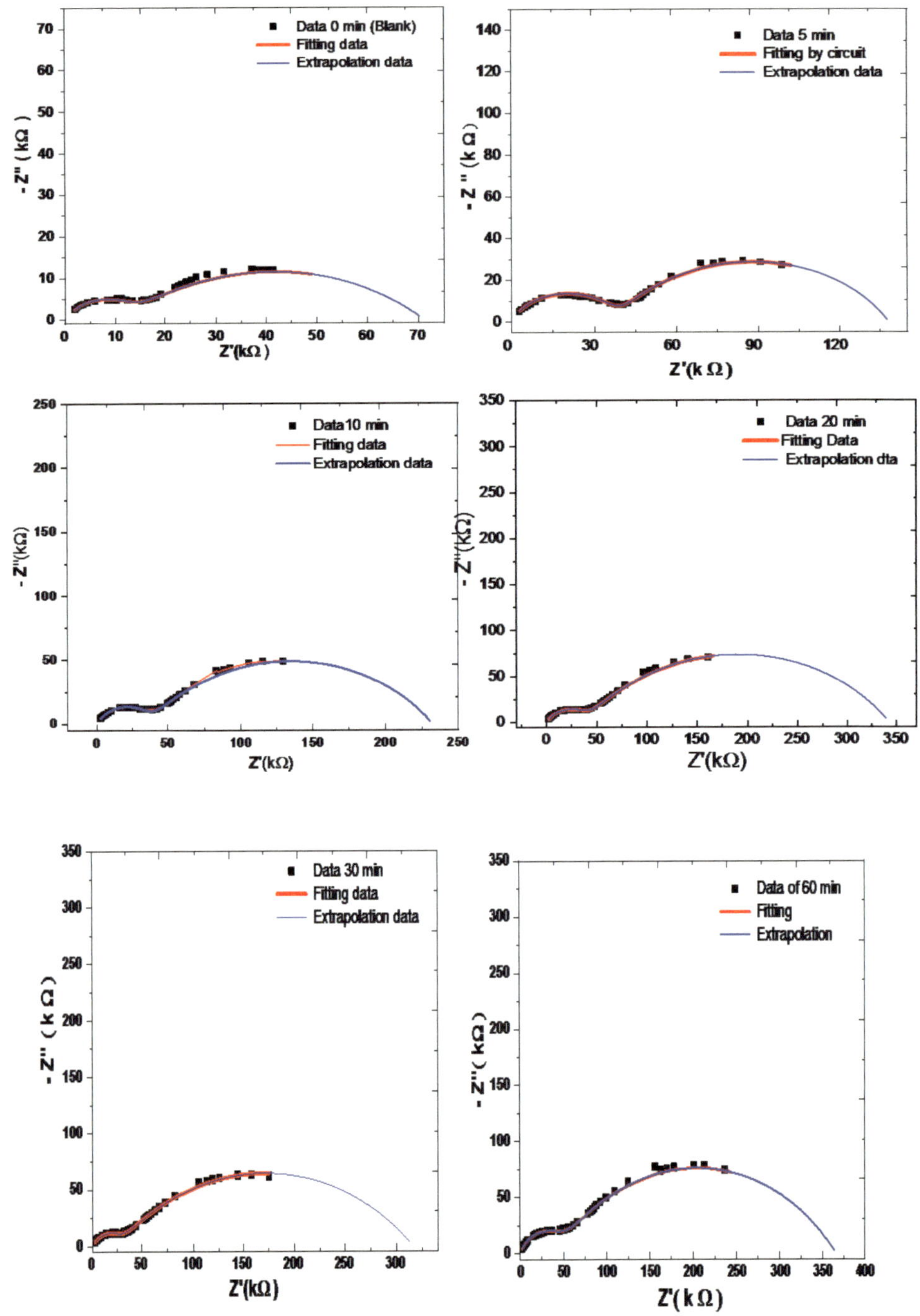

Figure 12.
Fitting and extrapolation of data experimental using equivalent circuit at all times of adsorption at room temperature.

the phenomenon of adsorption of the nitrate ions to the available pores of Zn_3-Al-Cl-LDH.

4.1 Intraparticle diffusion rate constant

In order to test the existence of intraparticle diffusion in the adsorption process, the amount of nitrate adsorbed per unit mass of adsorbents q at any time t was

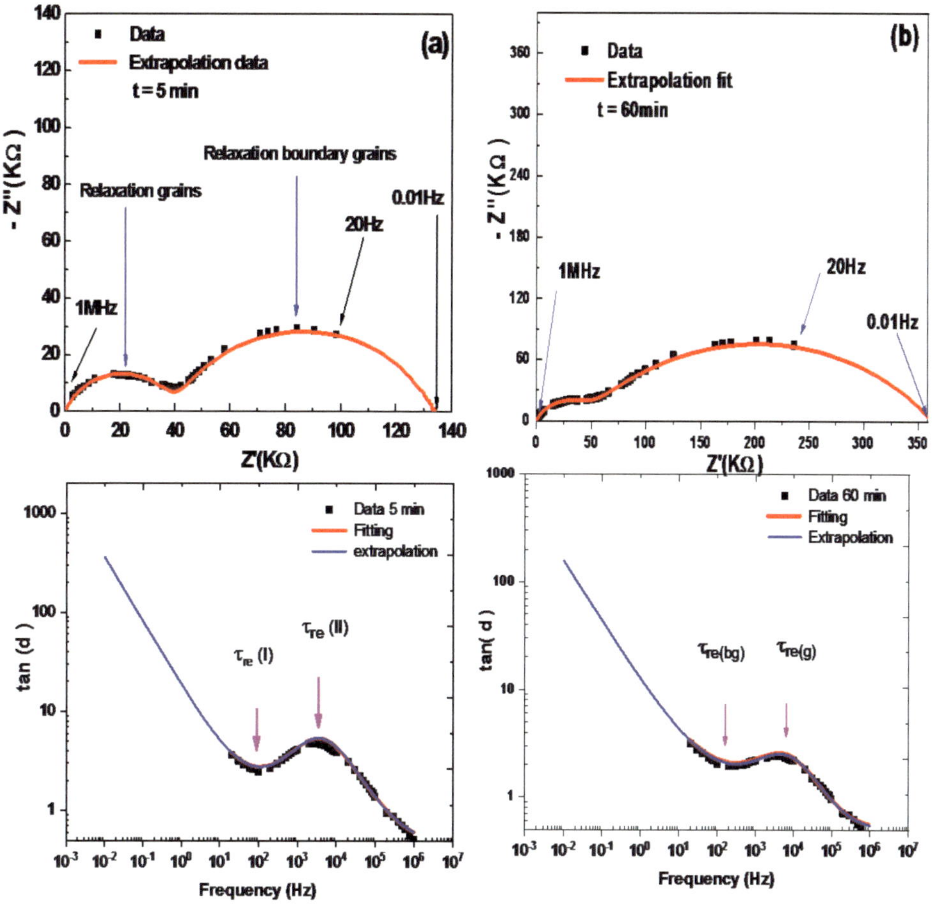

Figure 13.
Extrapolation of tan (d) measurement data of Zn_3-Al-Cl-LDH for t = 5 min (a) and (b) t = 60 min.

Sample	Rg (Ω)	(E%)g	R_{bg} (Ω)	$(E\%)_{bg}$	pg	p_{bg}	Tg ($\Omega^{-1}.s^{\alpha}$)	T_{bg} ($\Omega^{-1}.s^{\alpha}$)	X^2
0	11,513	0	61,702	0	0.71	0.43	2.05	1.42	0.003
5	36,160	2.14	157,410	1.55	0.73	0.63	1.90	0.34	0.001
10	40,275	2.49	196,670	2.19	0.74	0.62	1.64	0.23	0.001
20	41,753	2.62	302,160	3.90	0.75	0.63	1.49	0.28	0.001
30	42,584	2.69	298,590	3.84	0.76	0.64	1.43	0.24	0.0008
60	43,504	2.77	319,590	4.18	0.77	0.66	1.39	0.14	0.0001

Table 3.
The values of the fitted corresponding equivalent circuit parameters.

plotted as a function of square root of time ($t^{1/2}$). The rate constant for intraparticle diffusion was obtained using the Weber-Morris equation given as follows [35]:

$$q(t) = k_p.\sqrt{t} + c \quad (6)$$

where q is the amount of nitrate adsorbed in mg/g of adsorbent, k_p is the intraparticle diffusion rate constant, and "t" is the agitation time in minutes. Due to stirring, there is a possibility of transport of nitrate species from the bulk into the pores of the LDH as well as adsorption at an outer surface of the LDH. The rate-limiting step may be either adsorption or intraparticle diffusion.

Different regions of a system sample are characterized using a resistance and a constant phase element (CPE) placed usually in parallel, where subindexes "g" and "gb" refer to grain and grain boundary, respectively (Eqs. 7 and 8):

$$\tau_g = \left(R_g.T_g\right)^{\frac{1}{p_g}} \quad \text{and} \quad \tau_{jg} = \left(R_{jg}.T_{jg}\right)^{\frac{1}{p_{jg}}} \tag{7}$$

$$C_g = R^{\frac{1}{pg}-1}.T^{\frac{1}{pg}} \quad \text{and} \quad C_{jg} = R^{\frac{1}{pjg}-1}.T^{\frac{1}{pjg}} \tag{8}$$

The values of the individual $R_g.C_g$ and $R_{bg}.C_{bg}$ components may then be quantified. Let us now see some practical examples of data and interpretation. A common type of impedance spectrum for Zn_3-Al-Cl-$(NO_3^-)_{ads}$ LDH shows the presence of two distinct features attributable to intergrain or bulk and intergrain or grain boundary regions, using Eqs. (7) and (8) for obtained correspondent values listed in **Table 4**.

4.2 AC conductivity analysis

The ionic conductivities extracted from the data using the equivalent circuit of **Figure 14** depicted in **Figure 15a** and **b** show the variation of AC conductivity with frequency at various times of adsorption for nitrate ions in the surface of the ionic clay. The log–log curves are flat in the low-frequency region as the conductivity values approach those of the DC conductivity. As frequency increases, the curves become dispersive. In the high-frequency range, weak time dependence may be noted, and it is evident that the shapes of the curves are similar. In most materials AC conductivity due to localized states may be described using the equation of double power law of Jonscher [29]. In the electrical conductivity at different times of adsorption, it is clear from the plot that above a certain point, the conductivity increases linearly with frequency. From **Figure 8**, it is also evident that the DC contribution is important at low frequencies and the high time of adsorption, whereas the frequency-dependent term dominates at high frequencies [30–34].

It can be observed (**Figure 15a** and **b**) that increased with increasing frequency. This can be explained in terms of conductivity of grains separated by highly resistive grain boundaries. According to this model, the AC conductivity at low frequencies exhibited the grain boundary behavior, while the dispersion at high frequency is attributed to the conductivity of grains. This variation corresponding to the interpretation of LDH materials has two types of charge carrier, which are responsible for the dielectric relaxation [35]. As reported in our earlier article [36], the proton of the polarized clusters of water is the first carrier, and the nitrate ions

Time (min)	τ_g (ns)	τ_{jg} (μs)	C_g (pF)	C_{jg} (nF)	R_{jg}/C_{jg}	K_p
0	1.10	3.91	0.95	0.63	0.13	0.00
5	2.00	5.35	0.51	0.49	0.29	0.62
10	2.60	10.9	0.64	0.55	0.47	0.83
20	2.90	19.95	0.79	0.66	0.60	0.92
30	3.06	17.13	0.69	0.57	0.69	0.96
60	3.36	9.03	0.77	0.28	0.95	1.00

Table 4.
Results obtained by the corresponding equivalent electrical circuit.

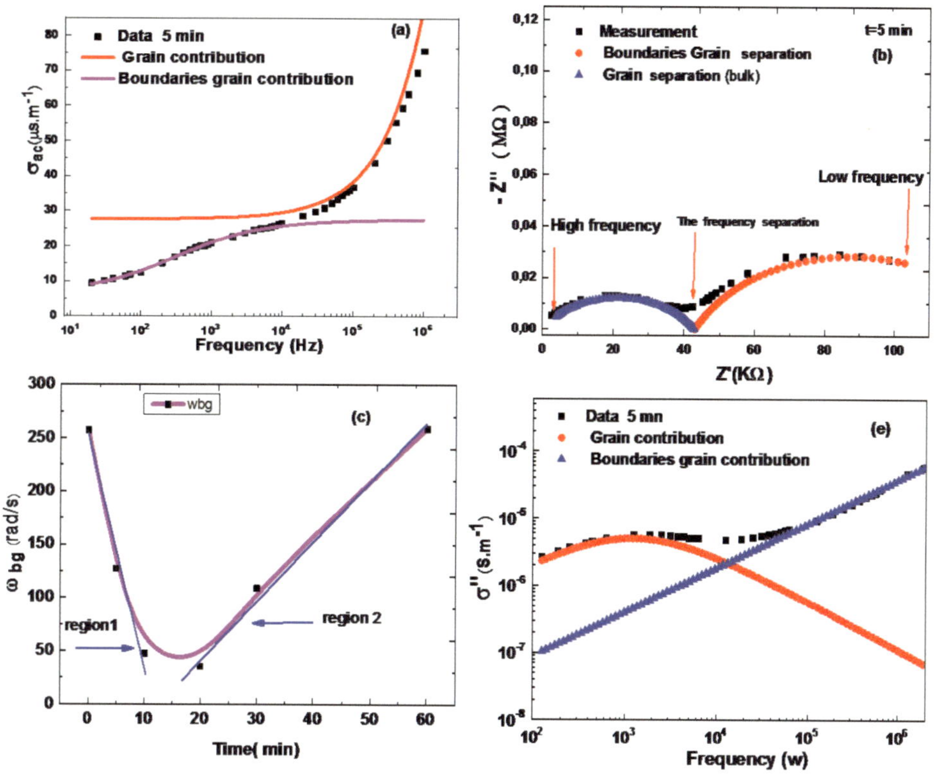

Figure 14.
Separation behavior of the conductivity (a) of grains and grain boundaries. (b) Nyquist diagram showing the behavior of the grain and grain boundary for the adsorption time 5 min. (e) and (e) in the representation of the imaginary part of the conductivity. (c) Variation of the grain boundary pulsation for the adsorption time 5 min.

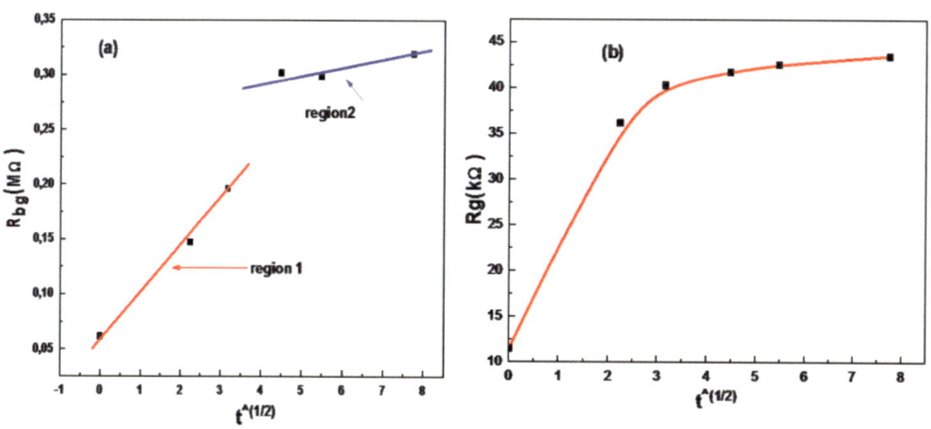

Figure 15.
Variation of the resistor (a) grain boundaries and (b) grains as a function of square root of time.

of the adsorption surface of LDH region (**Figure 2b**) is the second one. The proton transfers to produce OH^- and HO_3^+ ions (hydrogen bonding) at every path end of water clusters in the presence of the applied electric field due to proton hopping (inter-cluster hopping at low-frequency region and intra-cluster hopping at high frequency) (**Figure 2c**). In this condition, nitrate ions also transfer from their equilibrium positions to serve as an additional charge carrier.

Figures 13 and **15** show that σ'ac (ω) becomes almost independent of frequency below a certain value when it decreases with decreasing frequency. The ionic σ'ac (ω) conductivity will be obtained using the technical extrapolation of this part of spectra toward lower frequency.

The conductivity σ'ac frequency dependence can be described in the majority of ionic conductors by the simple power law Jonscher according to [37] descript be one term dispersion although our system provides else dispersion term depicted in **Figures** 7 and **8**.

$$\sigma'_{ac}(\omega) = \sigma_{DC} + A_g\omega^{p_g} + A_{bg}\omega^{p_{bg}} \tag{9}$$

The charge carriers are the adsorbed nitrate ions, and the protons originate from adsorbed mobile water located on the surface of the clay [4]. On the other hand the charge carriers are responsible for the second jump which is generated by the anions Cl^- and the H_3O^+ ions intercalated in the interlamellar region according to [38].

The slope changes in the conductivity variation depicted in **Figure 10a** and **b** confirms that the conductivity of our system exhibits two behaviors of frequency dispersion: low-frequency dispersion associated with grains and the other for grain boundaries (**Figure 14a** and **b**).

Usefulness of figures: **Figure 10a** of variation suggests the presence of a hopping mechanism in these samples. Such type of conducting behavior is well described by Jonscher's universal power law. On the other hand, our system presents two power laws.

The **Figure 10b** of variation suggests two contributions grains and boundaries grains. This manifest itself in the conductivity diagram, with two hopping conductions which lead to two different slopes.

The figure shows that the joint region of grains is the regions that adsorb nitrate ions, but this variation influences the grain region. On the other hand, this finding is in good agreement with the evolution shown in **Figure 16a**.

The figure presents the deconvolution in order to separate contributions grains and grain boundaries as a function of frequency.

4.3 Modeling of the electrical conductivity through the equivalent electrical circuit

From the modeling performed by an electrical circuit during the study of intercalation according to [39, 40], we have learned some physical characteristics of this

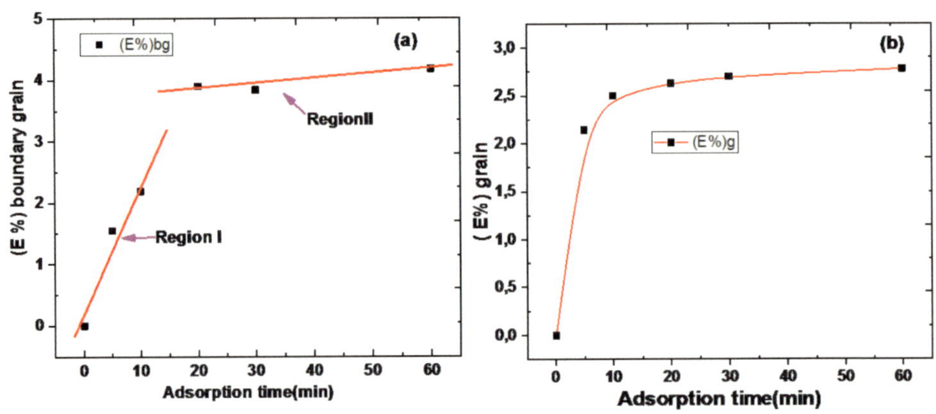

Figure 16.
Variation of the percentage of the relative difference in (a) grain boundary resistance and (b) grain at the different times of adsorption.

material. The circuit consists of a resistor in parallel with a constant phase element (CPE) characterized by a pseudocapacitance and Tp (scattering coefficient).

The conductivity of this circuit is of the following form [40]:

$$\sigma'_{ac}(\omega) = \frac{e}{S} \times y'(\omega) \tag{10}$$

where the real part of the admittance is

$$Y'(\omega) = \frac{1}{R}\left(1 + (\omega\tau)^{p} \cos\left(\frac{p\pi}{2}\right)\right) \tag{11}$$

Using the simulation of Eq. 11, we have obtained the equation of the admittance of grain established by the following expression:

$$\mathrm{Y}'_g(\omega) = \frac{1}{R}\left[1 + (\omega\tau)_g^{p_g} \cos\left(\frac{p_g\pi}{2}\right)\right] \tag{12}$$

Finally the conductivity of grain is shown in the following equation:

$$\begin{aligned} \sigma'_{ac}(\omega) &= \frac{k}{R}\left[1 + (\omega\tau)_g^{pg} \cos\left(\frac{pg\pi}{2}\right)\right] = \sigma_{dcg} + \frac{k}{R} \times (\omega\tau)_g^{pg} \cos\left(\frac{p_g\pi}{2}\right) \\ \sigma'_{ac}(\omega) &= \sigma_{dcg}\left(1 + (\tau)_g^{p_g} \cos\left(\frac{p_g\pi}{2}\right) \times \omega^{p_g}\right) \\ \sigma'_{ac}(\omega) &= \sigma_{dcg}\left(1 + \left(\frac{\omega}{\omega_{hg}}\right)^{pg}\right) \end{aligned} \tag{13}$$

with $\omega_{hg} = \left\{\tau_g^{pg} \cos\left(\frac{pg\pi}{2}\right)\right\}^{-1}$ the pulsation of hopping

In the term of conductivity, plots of the lower frequency dispersion correspond to the presence in grain boundary; this region can be approximately modeled by the circuit similar to the one depicted in **Figure 17** (grain boundary).

By similar calculations, the expression of the grain seal conductivity follows the shape of the following expression:

$$\begin{aligned} \left(\sigma'_{ac}(\omega)\right)_{bg} &= \frac{k}{R} \times \left[1 + (\omega\tau)_{bg}^{p_{bg}} \times \cos\left(\frac{pbg\pi}{2}\right)\right] \\ &= (\sigma_{dc})_{\mathrm{bg}} + \frac{k}{R} \times (\omega\tau)^{p_{gb}} \cos\left(\frac{p_{bg}\pi}{2}\right) \\ \left(\sigma'_{ac}(\omega)\right)_{\mathrm{bg}} &= (\sigma_{dc})_{\mathrm{bg}} \times \left[1 + (\tau)_{bg}^{p_{bg}} \cos\left(\frac{p_{bg}\pi}{2}\right)\right] \times \omega^{p_{bg}}, \\ \left(\sigma'_{ac}(\omega)\right)_{\mathrm{bg}} &= (\sigma_{dc})_{\mathrm{bg}} \times \left[1 + \left(\frac{\omega}{\omega_{hbg}}\right)^{p_{bg}}\right] \end{aligned} \tag{14}$$

with $\omega_{h_{bg}} = \left\{\tau_{bg}^{p_{bg}} \cos\left(\frac{p_{bg}\pi}{2}\right)\right\}^{-1}$ is the pulsation of hopping

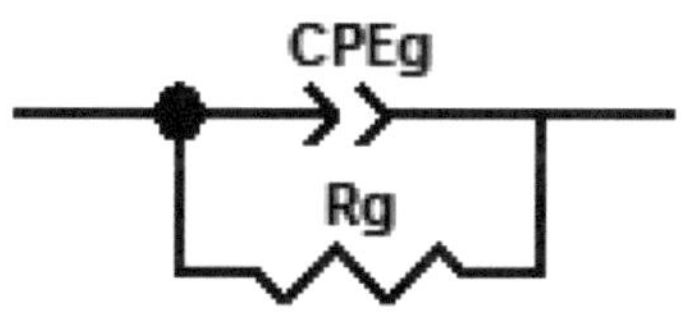

Figure 17.
Corresponding equivalent circuit used to fit by EIS experimental data of intercalation.

The total conductivity of the sample is similar for the different adsorption times; the evolution is typical called the double power law of Jonscher [41].

From the modeling performed by an electrical circuit during the adsorption study, we learned some physical characteristics of this material through the equivalent circuit modeling of the grain boundaries according to literature reviews [42] (**Figure 18**).

Figure 19 shows the electrical conductivity of the sample as a function of frequency. Two different regions can be distinguished. In Region 1, the conductivity is dominated by grains and in Region 2 is dominated by boundary grains [42], where the conductivity increases with increasing frequency. The electrical conduction of the sample follows a consecutive hopping mechanism. Whenever it is transferred to another site, the surrounding molecules respond to this perturbation with structural changes, and the electron or hole is temporarily trapped in the potential well leading to polarization. Another aspect of this charge hopping mechanism is that the electron or hole tends to associate with local defects [3, 43].

The dependence of the AC conductivity on frequency can be expressed by the following law:

$$\sigma'_{ac}(\omega) = \sigma_{DC} + A_g \times \omega^{p_g} + A_{bg} \times \omega^{p_{bg}} \tag{15}$$

where Ag is a pre-exponential factor and Abg is the frequency exponent [8, 44–46], which generally is less than or equal to 1. **Figures 7** and **8** show the frequency dependence of the AC electrical conductivity at different times of adsorption. It is clear from the plot that above a certain point, the conductivity increases linearly with frequency. In these figures it is also evident that the DC contribution is important at small frequencies and high frequencies, whereas the frequency-dependent term dominates at high frequencies. Also in the low-frequency region, the conductivity depends on the time of adsorption. Such dependence may be described by the variable range hopping (VRH) mechanism also called hopping conduction mechanism. The value of A_g and A_{bg} in Eq. (13) was extracted from the slope of the plot of Log(σ'ac) versus Log(f), and this value was used to explain the conduction mechanism in the sample. The capacitance adsorption is called double layer capacitance is the dependence of values was plotted, and it is seen that the frequency

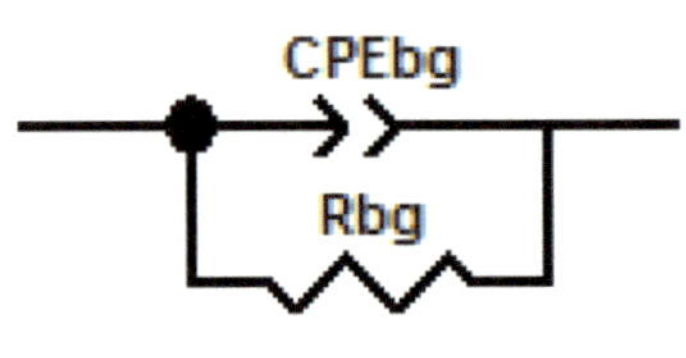

Figure 18.
Corresponding equivalent circuit to fit by EIS experimental data contribution of cluster (NO_3^- H_3O^+) adsorbed on the surface (grain boundary).

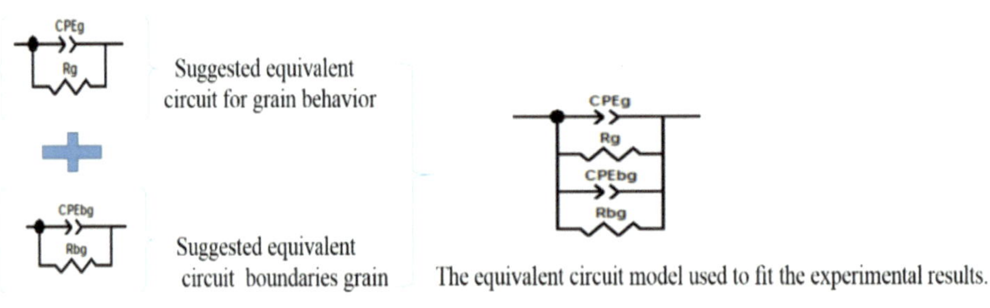

Figure 19.
Corresponding equivalent circuit used to fit the EIS experimental data of conductivity for sample study.

exponent decreases with increasing surface adsorption. This result is in clear agreement with the correlated barrier hopping (CBH) model, so the frequency dependence of σ'ac can be explained in terms of this model [47–53].

Better fundamental understanding of the adsorption phenomenon is modeled by the equivalent circuit depicted in parentheses (**Figure 19**).

This circuit has the expression of conductivity is following Eqs. 10. Use the fit by this circuit the results of the simulation are tabulated in **Table 5**. The values of the conductivity of grains and grain boundaries fitted and evaluated using the equivalent electrical circuit are reported in **Table 5**.

All frequency sweeps from this experiment were analyzed using the model of double power law to give σbg values σ_g grain relaxation time and grain boundary. In all cases the double power law provided an excellent fit to the data; **Figure 10a** and **b** is a plot of σ'_{dc} values as a function of frequency for the two regions.

In the model hopping, we distinguish two different characteristics in measurement frequency range. The charge transport takes place via an infinite percolation path in intermediate frequency. At high frequencies when the conductivity increases, the transport is dominated by a hopping contribution in finished areas of the system and is manifested in the variation of conductivity as a function of frequency by slope breaks such as p_g and p_{bg}.

4.4 Correlation between kinetic and impedance spectroscopy studies

Throughout in this study, the electrical properties of adsorption of nitrate by system LDH using the spectroscopy impedance as technical the investigation and monitoring of adsorption is important for the excellent result using the resistor of grains and the resistor of boundaries grains (**Figure 20**).

In order to attributed the second semicircle to a feature of the system, it is essential to have a picture of an idealized $Zn_3Al\text{-}Cl\text{-}(NO_3^-)_{ads}$-LDH with grains and grain boundaries and consider the factors which control the magnitude of the grain boundary impedance. **Figure 2b** model represents a $Zn_3Al\text{-}Cl\text{-}(NO_3^-)_{ads}$-LDH.

These results such as those shown in **Figure 21a-d** are useful for several reasons:

- To indicate whether the overall resistance of a material is dominated by bulk or grain boundary assured by the adsorption of nitrate
- To assess the quality and electrical homogeneity in the monitoring of adsorption since there is generally a link between sintering/microstructure and AC response.
- To measure the values of the resistances and capacitances at different times of adsorption

Time (min)	σ_g (μS/m)	σ_{bg} (μS/m)	τ_{bg} (μS)	τ_g (ms)	C_{bg} (nF)	C_g (pF)
0	8.685	1.620	1.10	3.89	1.01	0.634
5	2.770	0.7812	1.69	7.86	0.601	0.713
10	2.95	0.369	2.02	21.16	0.752	0.784
20	3.044	0.2493	2.58	28.26	0.983	0.704
30	3.761	0.34602	2.85	7.2	0.795	0.550
60	2.29	0.3125	3.35	3.86	0.895	0.121

Table 5.
Values of the conductivity relaxation time and the capacitance of grains and grain boundaries.

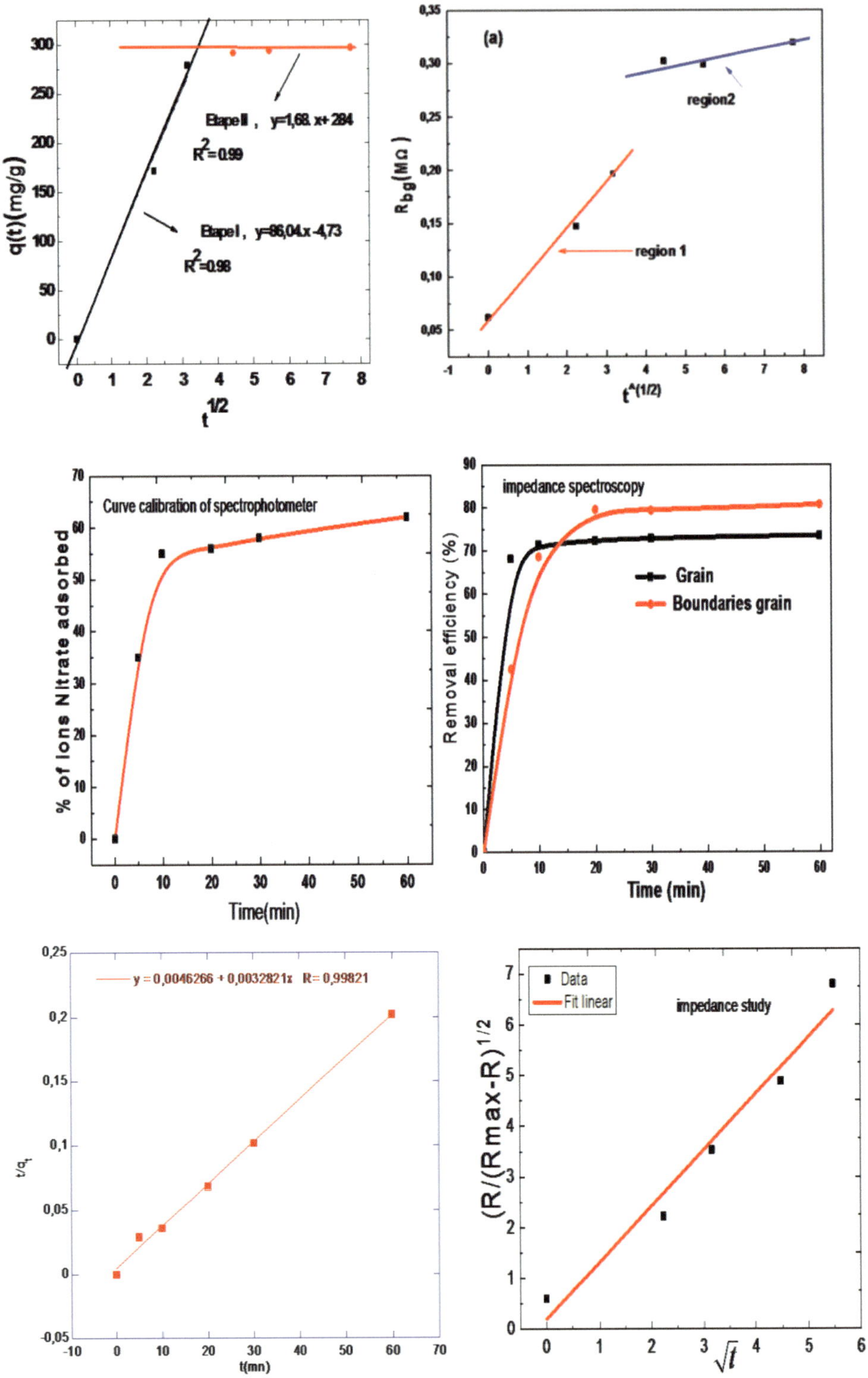

Figure 20.
Correlation figures between the kinetic and impedance spectroscopy studies of adsorption.

The distinction between kinetic and complex impedance spectroscopy study later controls the magnitude of the grain and the grain boundary and with a typical bulk permittivity in the range 103 to 105 using the relation and

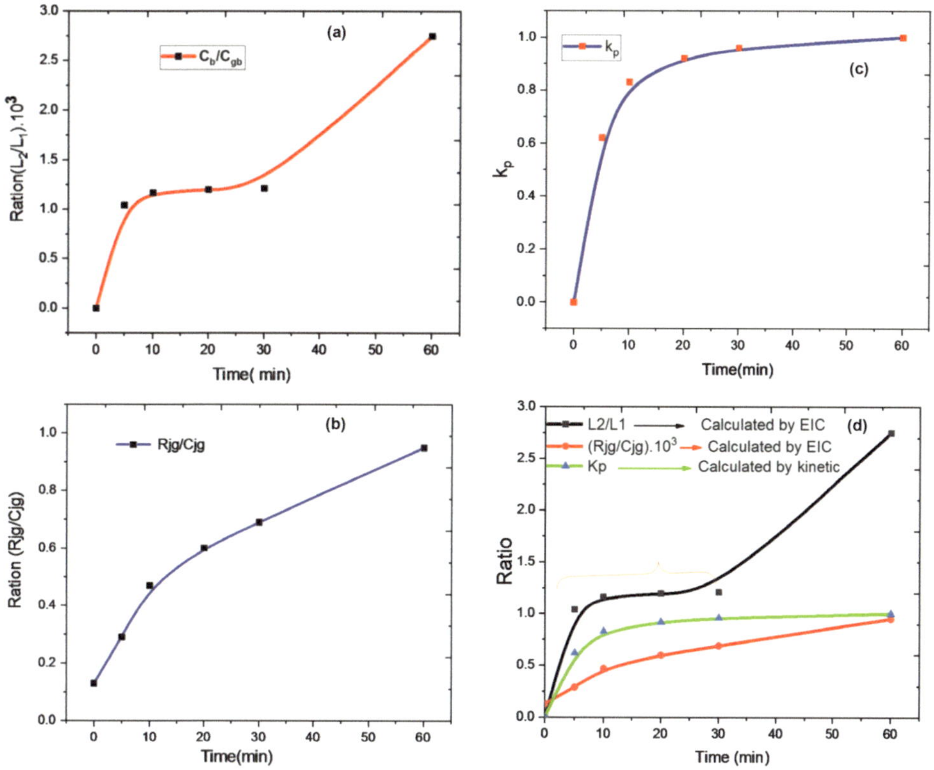

Figure 21.
(a) Ratio dimensional between grain and grain boundaries as a function of adsorption time. (b) Variation the fractality of system versus adsorption time. (c) Evolution of the partition coefficient between absorbance and absorbance as a function of time. (d) Variation in a comparative study between impedance spectroscopy and the kinetic study indicating the adsorption equilibrium.

visualized in **Figure 17** show the variation of imaginary part of permittivity at 1 kHz the frequency.

5. Conclusion

Characteristic band of NO_3^- and LDH, which indicated to the intercalating process, has been successful in the FT-IR diagram. Dielectric response of Zn_3-Al-Cl-LDH samples has been explained using the Cole-Cole presentation during the adsorption phenomenon. The resistor of the sample increased from 73,215 to 363,094 Kohms, and also the conductivity spectra exhibited high conductivity in high frequency according to two mechanisms of hopping conduction: one of the water molecules and the other of the nitrate ions adsorbed in the LDH. The mathematical fitting obtained using the equivalent circuit of these diagrams was carried out to obtain the conductivity following the double power law of Jonscher.

Acknowledgements

The author would like to thank the reviewers for their insight and their painstaking.

Funding statement

This research did not receive any specific grant from funding agencies in the public, commercial or not-for-profit sectors.

Competing interest statement

The authors declare no conflict of interest.

Additional information

No additional information is available for this paper.

Author details

Abderrahmane Elmelouky*, Abdelhadi Mortadi, Elghaouti Chahid and Reddad Elmoznine
Laboratory Physics of Condensed Matter, University Chouaib Doukkali, El Jadida, Morocco

*Address all correspondence to: elmlouky_abderrahmane@yahoo.fr

References

[1] Jeong JY, Kim HK, Kim JH, Park JY. Electrochemical removal of nitrate using ZVI packed bed bipolar electrolytic cell. Chemosphere. 2012;**89**:172-178. DOI: 10.1016/j.chemosphere.2012.05.104

[2] Zhou M, Fu W, Gu H, Lei L. Nitrate removal from groundwater by a novel three-dimensional electrode biofilm reactor. Electrochimica Acta. 2007;**52**:6052-6059. DOI: 10.1016/j.electacta.2007.04.010

[3] Thomson TS. Nitrate concentration in private rural drinking water supplies in saska Tchewan Canada. Bulletin of Environmental Contamination and Toxicology. 2001;**66**:64-70. DOI: 10.1007/s001280000206

[4] Majumdar D, Gupta N. Nitrate pollution of groundwater and associated human health disorders. Indian Journal of Environmental Health. 2000;**42**:28-39. DOI: 10.4236/jeas.2015.54015

[5] Tate CH, Arnold KF. Health and aesthetic aspects of water quality. In: Pontius FW, editor. Water Quality and Treatment. New York: McGraw-Hill Inc.; 1990. pp. 63-156

[6] Zheng Y, Wang A. Nitrate adsorption using poly (dimethyl diallyl ammonium chloride)/polyacryl amide hydrogels. Journal of Chemical & Engineering Data. 2010;**55**:3494-3500. DOI: 10.1021/je100169r

[7] Chatterjee S, Woo SH. Removal of nitrate from aqueous solutions by chitosan hydrogel beads. Journal of Hazardous Materials. 2008;**164**:1012-1018. DOI: 10.1016/j.jhazmat.2008.09.001

[8] Reichel WT. Synthesis of anionic clay minerals. Solid State Ionics. 1986;**22**:135-142. DOI: 10.1016/0167-2738 86) 90067-6

[9] El Ouardi M, Qourzal S, Alahiane S, Assabbane A, Douch J. Effective removal of nitrates ions from aqueous solution using new clay as potential low-cost adsorbent. Journal of Encapsulation and Adsorption Sciences. 2015;**05**(04):178-190. DOI: 10.4236/jeas.2015.54015

[10] Wakelin SA, Colloff MJ, Kookanal RS. Effect of wastewater treatment plant effluent on microbial function and community structure in the sediment of a freshwater stream with variable seasonal flow. Applied and Environmental Microbiology. 2008;**74**:2659-2668. DOI: 10.1128/AEM.02348-07

[11] Dabrowski A. Adsorption—From theory to practice. Advances in Colloid and Interface Science. 2001;**93**:135-224

[12] Bates JB, Chu YT, Stribling WT. Surface topography and impedance of metal electrolyte interfaces. Physical Review Letters. 1988;**60**:627

[13] Horvat-Radosevic V, Kvastek K. Hydrogen/anion electro sorption at rhodized electrodes as revealed by electrochemical impedance spectroscopy. Journal of Electroanalytical Chemistry. 2004;**566**:451-465

[14] Yoo HD, Jang JH, Ka BH, Rhee CK, Oh SM. Impedance analysis for hydrogen adsorption pseudo capacitance and electrochemically active surface area of Pt electrode. Langmuir. 2009;**25**(19):11947-11954

[15] Barbero G. Influence of adsorption phenomena on the impedance spectroscopy of a cell of liquid. Physical Review E. 2005;**71**:062201

[16] Barbero G, Becchi M, Strigazzi A, Digabel JL, Neto AMF. Experimental evidence for the adsorption-desorption phenomena on the spectroscopy impedance measurements of an

electrolytic cell. Journal of Applied Physics. 2007;**101**:044102

[17] Bataliotoa F, Martins OG, Duarte AR, Neto AMF. The impudence of adsorption phenomena on the impedance spectroscopy of an electrolytic cell. European Physical Journal E: Soft Matter and Biological Physics. 2011;**34**:10

[18] Nakanishi M. Graphical conversion between compliance and modulus, permittivity and electric modulus, impedance and admittance. International Journal of Spectroscopy. 2014:7. DOI: 10.1155/2014/538206. Article ID 538206

[19] El Melouky A, El Moznine R, Lahkale R, Sadik R, Sabbar E, Chahid E, et al. Journal of Optoelectronics and Advanced Materials. 2013:1239-1247

[20] Ahmed AAA, Talib ZA, Zobir Bin Hussein M, Zakaria A. Solid State Sciences. 2012:1196-1202

[21] Yadav AK, Akashi CP, Haritash AK, Kansal A, Rani N. Journal of Hazardous Materials. 2006;**128**:289-293

[22] Yang L, Maomao Yang ID, Xu P, Zhao X, Bai H, Li H. Characteristics of nitrate removal from aqueous solution by modified steel slag. Water. 2017;**9**: 757. DOI: 10.3390/w9100757

[23] Ahmed AAA, Talib ZA, Zobir Bin Hussein M, Zakaria A. Journal of Solid State Chemistry. 2012;**191**:271-278

[24] Reinhardt C. Weinheim (Federal) Germany: VCH Verlags y Eseuscha. D-6940. pp. 75–78

[25] Lagergren S. ZurTheorie der Sogenannten Adsorption Gelöster Stoffe. Kungliga Svenska Vetenskap sakademiens. Handling 1898;**24**:1–39 (IGermany)

[26] Ho YS, Mckay G. Sorption of dye from aqueous solution by peat. Chemical Engineering Journal. 1998;**70**: 115-124

[27] Hamou RF, Macdonald JR, Tuncer E. Dispersive dielectric and conductive effects in 2D resistor-capacitor networks. Journal of Physics: Condensed Matter. 2009;**21**:025904

[28] Randles JEB, Somerton KW. Transactions of the Faraday Society. 1952;**48**:937-950

[29] Zhu X-h, Li J, Jin Y, Guo Y-h. Preparation of porous hybrid adsorbents based on fluor(calcium silicate)/ activated carbon and its application in the removal of iron (III) from ammonium phosphate solutions. Arabian Journal of Chemistry. 2018. DOI: 10.1016/j.arabjc.2017.12.006

[30] Funke K. Solid State Chemistry. 2007;**22**:111

[31] K. Zhao. Chemical Industry Press. Beijing. 2008

[32] Lv L, He J, Wei M, Evans DG, Duan X. Water Research. 2006;**40**(4): 735-743

[33] Abe M, Ogino K. Journal of Colloid and Interface Science. 1981;**80**:58-66

[34] Ismaïli M, Bougroua F, Isaert N, Legrand C, Nguyen HT. Physical Review E. 2001;**65**:011701

[35] Islam M, Patel R. Synthesis and physicochemical characterization of Zn/Al chloride layered double hydroxide and evaluation of its nitrate removal efficiency. Desalination. 2010;**256**:120

[36] Senthil Kumar P, Sakunthala A, Prabu M. Impact of cerium doping on the structural and electrical properties of lithium nickel manganese oxide ($LiNi_{0.5}Mn_{0.5}O_2$). International Journal of ChemTech Research. 2014;**6**: 5252-5255

[37] Córdoba-Torres P. Characterization of frequency dispersion in the impedance response of a distributed model from the mathematical properties of the distribution function of relaxation times. Electrochimica Acta. 2015;**180**: 591

[38] Liang P, Yang Z, Chao X. Journal of Alloys and Compounds. 2016;**678**: 273-283

[39] El Moznine R, Smith G, Polygalov E, Suherman PM, Broadhead J. Journal of Applied Physics. 2003;**36**:330-333

[40] Roto R, Villemure G. Journal of Electroanalytical Chemistry. 2002;**527**: 123-130

[41] Gupta SS, Bhattacharyya KG. Advances in Colloid and Interface Science. 2011;**162**:39-58

[42] Badapanda T, Senthil V, Rout SK, Cavalcante LS, Simões AZ, Sinha TP, et al. Current Applied Physics. 2011;**11**: 1282-1293

[43] Hajra S, Sahoo S, Mishra T, De M, Rout PK, Choudhary RNP. Journal of Materials Science: Materials in Electronics. 2018;**29**:7876-7884. DOI: 10.1007/s10854-018-8787-8

[44] Mehrotra V, Giannelis EP. Journal of Applied Physics. 1992;**72**:1039-1048

[45] Elhatimi W, Lahkale R, Bouragba FZ, Sadik R, Lebbar N, Elmelouky A, et al. Journal of Spectroscopy Letters. 2017

[46] Liang X, Hou W, Xu Y, Sun G, Wang L, Sun Y. Physicochemical and Engineering Aspects. 2010;**366**:50-57

[47] Ho YS. Journal of Hazardous Materials. 2006;**136**:681-689

[48] Elmelouky A, Chahid E, Elmoznine R, Lahkale R, Sabbar E. Analysis of conduction mechanism in lamellar double hydroxide by impedance spectroscopy. Journal of Optoelectronics and Advanced Materials. 2013;**15**(11):1239-1247

[49] Ahmed AAA, Talib ZA, Bin Hussein MZ, Zakaria A. Influence of metallic molar ratio on the electron spin resonance and thermal diffusivity of ZnRAl layered double hydroxide. Journal of Solid State Chemistry. 2013: 271-278

[50] Liang PF, Zhu T, Fang Y, Li Y, Han Y, Wu Y, et al. Atmospheric Chemistry and Physics. 2017;**17**:13921-13940

[51] Lahkale R, Elhatimi W, Sadik R, Bouragba FZ, Lebbar N, El A, et al. Nitrate-intercalated $Mg_{1-x}Al_x$-layered double hydroxides with different layer charges (x): Preparation, characterization, and study by impedance spectroscopy. Applied Clay Science. 2018;**158**:55-64

[52] Elhatimi W, Bouragba FZ, Lahkale R, Sadik R, Lebbar N, Siniti M, et al. Solid State Sciences. DOI: 10.1016/j.solidstatesciences.2018.03.006

[53] Elmelouky A, Mortadi A, Chahid E, Elmoznine R. Impedance spectroscopy as a tool to monitor the adsorption and removal of nitrate ions from aqueous solution using zinc aluminum chloride anionic clay. Heliyon. 2018;**4**. DOI: 10.1016/j.heliyon.2018.e00536

Chapter 8

Electrochemical Impedance Spectroscopy (EIS) in Food, Water, and Drug Analyses: Recent Advances and Applications

Marwa El-Azazy

Abstract

Electrochemical impedance spectroscopy (EIS) is a *potent* electrochemical technique with a variety of applications. EIS measurements involve the application of an alternating current (AC) voltage (or current) to the system under investigation, followed by measurement of the response in the form of AC current (or voltage) as a function of frequency. By and large, EIS is an exceptionally attractive in terms of applications. Being nondestructive with a feasibility of implementation to the system to be measured and the usefulness of data obtained in characterizing the studied systems, electrochemical impedance spectroscopy has realms of applications. As food and water safety and security is becoming a universal concern, the need for a technique that can detect water and food contaminants with relatively high sensitivity and selectivity is evolving. EIS has started to realize its potential with a wide-term use in water and food analyses.

Keywords: electrochemical impedance spectroscopy, electrochemical techniques, food analysis, drug analysis, water analysis

1. Introduction

Electrochemical impedance spectroscopy (EIS) is a usually described as a *potent* (if not the most powerful) electrochemical analytical technique. The history of EIS goes back to the late nineteenth century, thanks to the foundations established by Heaviside on his work on the linear systems theory (LST). By the end of the same century, the success achieved by Warburg to broaden the conception of *impedance* to the electrochemical systems (ES) came to the scene. It was close to the middle of the twentieth century, when the EIS started to realize its potential! That came with the invention of the potentiostat in the 1940s, followed by the frequency response analyzers in the 1970s. This progress has led to the application of EIS chiefly in investigation of corrosion mechanisms [1–3].

Later on, this has opened the doors for realms of applications of EIS. Applications encompassed electrocatalysis and energy [3–5]; characterization of materials, e.g. corrosion phenomenon surveillance [6, 7]; and depiction of quality of coatings [8], exploring mechanisms of processes such as electrodeposition and electro-dissolution [9, 10], food and drug analysis [11–13], detection of biomarkers [14, 15], and water analysis [16, 17].

It is noteworthy to mention that impedance spectroscopy (IS), depending on the material used, the device, and the system or process to be studied, has two main categories: EIS (the topic of this chapter) and dielectric IS. A major difference is that EIS applies to systems/materials involving chiefly ionic conduction, in contrast to electronic conduction in the case of dielectric IS. Therefore, it can be observed from the fields of EIS applications that EIS usually applies to systems like electrolytes (solid/liquid), polymers, and glasses [18–21].

In general, EIS measurements involve the application of an alternating current (AC) voltage or current to the system under investigation, followed by measurement of the response in the form of AC current (or voltage) as a function of frequency. Measurements are usually performed using the potentiostat, and the measured response is analyzed using a frequency response analyzer (FRA) [18]. By and large, three factors make EIS exceptionally attractive in terms of applications:

1. Capability to explore the ES at relatively low frequencies using the minimal perturbation that in turn serves to keep the kinetic information of the system under investigation at near zero conditions (steady state). Therefore, EIS is said to be a steady-state and nondestructive technique. The majority of the electrochemical techniques, however, involve an application of large perturbation for sensing the membrane/electrolyte interface, with the purpose of obtaining mechanistic data following the driving of the reaction to a state that is far from equilibrium [3].

2. Feasibility of implementation of EIS into the system to be measured.

3. The usefulness of data obtained in characterizing the studied ES, where EIS provides on-site data on the relaxation data over a range of frequencies, from as low as 10^{-4} Hz and up to 10^{6} Hz.

A combination of the three advantages led to the wide use of EIS as previously mentioned.

The current chapter throughout the following sections is investigating the applications of EIS in a variety of matrices, mainly in food, drug, and water analysis, and the recent advances in these fields as well as comparisons between EIS and other electroanalytical approaches applied for the same purposes.

2. Chapter taxonomy

Throughout the current chapter, readers will be exposed to the different analytical techniques, especially the electrochemical-based approaches, which are generally used for detection of pollutants in food, drug, and water.

Readers will be more focused on the applications of EIS in specific. A comparison between EIS and the other techniques commonly used in water and food analysis will be exhibited.

3. Food and water security: the global concern

The safety, quantity, and the quality of food and water are becoming worldwide concerns. Water is the most crucial source for human development. With the advancement of human life, uncountable contaminants are intimidating the aquatic system. These intimidations include but not limited to automation/

industrialization, widespread use of chemicals, increased population, and suburbanization. Subsequently, water pollution is becoming a significant health and environmental concern.

By and large, the safety of food and water is influenced by contaminants. Among these pollutants, heavy metals, elevated anions (sulfates, phosphates, fluoride, etc.), dyes, agricultural waste, pesticides, drugs, and pharmaceuticals are the most common. Heavy metals, in specific, are widely used in many industrial, domestic, and agricultural applications, and they are nondegradable, an issue that raises the concern about their potential influence on public health, water systems, and the ecosystem in general. As, Cd, Cr, Pb, and Hg have been reported to be the highest systemic toxicant elements.

According to the US EPA and the International Agency for Research on Cancer (IARC), these toxic elements are probable to be carcinogenic. Moreover, accumulation of Pb, Cd, and Hg in the human body over time can cause serious health problems [22–26].

Similarly, food, leather, and textile industries discharge huge amounts of polluted wastes. With the various structures of the chemicals used in these industries, numerous problems develop. Dyes, a key water pollutant and even if discharged as traces (as low as 1 ppm), would color large volumes of water. Reports show that amount of dyes as huge as 7×10^5 tons per annum are being produced annually, demonstrating the magnitude of the problem. Released dyes do not only affect the aquatic beings but also the human health. Their impact includes carcinogenicity, mutagenicity, poisoning, disturbance of the metabolism in aquatic bodies, etc. [27, 28].

On the other hand, and representing a significant category of aquatic pollutants, pharmaceutically active materials (PhAMs) are usually released into the aquatic systems from different sources, including but not limited to the effluents of the manufacturing sites and hospitals, illegal disposal, veterinary applications, and landfill leachate. Daily use by humans and the subsequent conversion of PhAMs into various metabolites with variable chemical structures is also a major source. The fate of these metabolites, and probably their parent drug compound, is usually the wastewater [29–33].

The increasing understanding of the assembly of the food chain and the probability of infection of human with these resilient microorganisms, either directly or via the food chain, has explained largely the spread of these species. Therefore, the process of food production and commercialization is posing rigorous regulations nowadays. Different societies, such as the Food and Drug Administration (US FDA), European Union (EU), and World Health Organization (WHO) in collaboration with the Food and Agriculture Organization of the United Nations (FAO) creating the FAO/WHO Codex Alimentarius Commission (CAC), are setting up standards for the maximum residue levels (MRLs) permissible in raw and processed food products of animal or poultry origin. Yet, any food product that would conform to these criteria and the preceding risk assessments cannot be banned by countries of the World Trade Organization (WTO) [34–38].

The elevating concerns on food and water safety and the existence of these materials at relatively low concentrations have created the need to find sturdy as well as sensitive detection and removal/remediation technologies. Detection technologies included traditional techniques such as spectrophotometry, spectrofluorimetry, atomic absorption spectrometry (AAS), as well as electrochemical and the more sophisticated chromatographic approaches [27, 28, 39–49]. Each of these techniques has its pros and cons.

Electrochemical techniques are among the widely used techniques for detection of pollutants in food and water analyses. The following subsections will be focused on the electrochemical approaches and EIS in specific in detection of contaminants in water and food samples.

3.1 Electrochemical techniques: principles, advantages, and sensing mechanisms

As an analytical approach, electroanalysis offers many advantages including but not limited to simplicity, sensitivity, specificity, and applicability in various matrices and cost-effectiveness. These advantages are of specific importance when it comes to detection of drugs and pharmaceuticals, especially in food and water samples as well as in quality control (QC) and quality assurance (QA) laboratories. According to the signal being measured (voltage/potential, current, conductance, impedance), electroanalytical techniques can be categorized into potentiometric, amperometric, conductometric, and impedimetric techniques. Subcategories for each technique also exist, and coupling with other technologies has been investigated.

Sensors, and in particular those based on the classical potentiometric technique, or the new polyion, galvanostatic, or voltammetric sensing mechanisms, now possess the foothold in analytical chemistry. Offering irresistible advantages, on the in vitro scale, such as operation simplicity, sturdiness, and remarkable sensitivity hitting nine orders of magnitude, selectivity, and functionality over wide range of matrices, suitability for on-line or real-time analyses, and most prominently their liability for miniaturization, make the use of sensors indispensable [50–53].

Figure 1 shows an illustration of ISE (ion-selective electrode) potentiometric sensor and generation of potential across the different phase boundaries.

The sensing mechanism especially if the target analyte is a biomolecule depends on tailoring the surface of the sensor with a bio-receptor that can selectively bind to the target bio-analyte. Following the adsorption of the bio-analyte from the solution on the surface of the probe, a change in the electrochemical signal can be observed and measured. Such a change is correspondingly dependent on the bio-analyte concentration.

Figure 2 shows a schematic illustration of the sensing mechanism in plastic microfluidic channels. The left panel shows the generation of streaming potential,

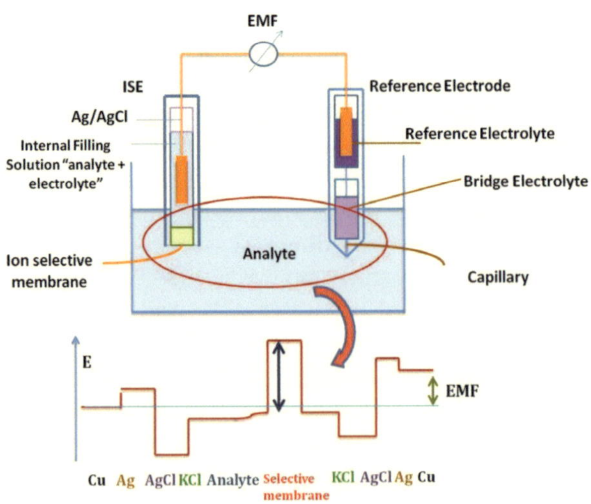

Figure 1.
Schematic illustration of ISE cell assembly and the generation of EMF across different phase boundaries.

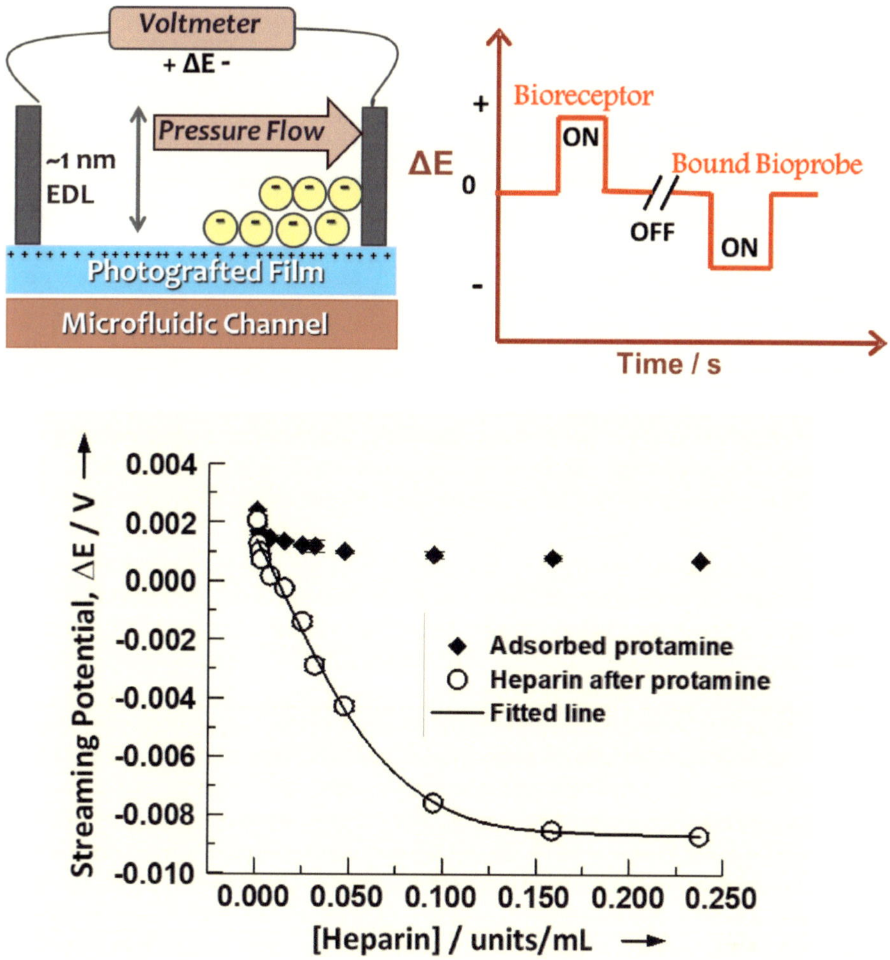

Figure 2.
Schematic illustration of the generation of streaming potential as a result of pressure-driven flow and surface charge at the electric double layer (EDL)—left upper panel. The right panel shows a sensogram with signal inversion upon adsorption of the analyte. The bottom graph shows the pulsed streaming potentials as a function of heparin with immobilized protamine on a surface of a cyclo-olefin copolymer (COC) microchannel. Data points were fitted using Langmuir isotherm. Graphs are replicated from the authors' own work with permission from Copyrights@ American Chemical Society (ACS) [45].

ΔE, as a result of pressure-driven flow and surface charge at the electric double layer (EDL). The right panel shows a sensogram with signal inversion upon adsorption of the analyte. The bottom graph shows the pulsed streaming potentials as a function of heparin with immobilized protamine.

EIS as an electrochemical technique entails measurement of the change in the charge transfer resistance (R_{ct}) following the interactions between the analyte and the receptor and the consequent change in the interfacial electron transfer kinetics. The following sections will be dealing with the application for EIS for sensing different target analytes in different matrices [53, 54].

4. EIS in drug analysis

The effects of presence of the PhAMs either in waste and drinking water or even in wastewater treatment plants (WWTPs) are still inarticulate. However, what is well understood is that the impact extends to humans and animal's health, the aquatic environment, and in the long run the ecosystem. This effect is greatly dependent on the released dose of the PhAMs as well as their pharmacological effects. The issue becomes of concern when we know that the metabolites might be of a higher risk than the parent drug compound. At the microbial level, microorganisms upon prolonged exposure to anti-infectives, for example, become more tolerant, and new strains, which cannot be cured using the conventional antimicrobials, are now in the scene [55–57].

EIS, being capable of detecting as low as 10^{-12} M of the target analyte, is widely used in drug analysis. Several drug categories were analyzed using EIS. **Table 1** shows some examples of drugs analyzed using EIS, as well as the matrices and type of electrode used together with the sensing interface, sensing strategy (label-free or labelled), and limit of detection (LoD).

The electrochemical properties of raloxifene, an important chemotherapeutic agent, were assessed using different techniques including EIS. Three electrodes were tested for this investigation: (1) bare screen-printed carbon electrode (SPCE), (2) graphene oxide (GO)/glassy carbon electrode (GCE), and (3) neodymium sesquioxide nanoparticles Nd_2O_5 NPs@GO/GCE. The target was to assess the interface properties of these electrodes. Results showed that the R_{ct} of the third electrode was much smaller than the other electrodes. Other electrochemical techniques such as cyclic voltammetry (CV) were used in the same work [58].

Other examples included the determination of an important class of PhAMs, which is antibiotics, a subclass of antimicrobials. Label-free detection of oxytetracycline (OTC) in milk samples was performed using a mixture of iron oxide and mesoporous carbon (Fe_3O_4@mC) together with nanocomposites made of Fe(II)-based metal-organic frameworks (525-MOF) by calcination at different temperatures. The sensor showed a very high sensitivity with a LoD = 0.027 pg mL^{-1} and a linear range of 0.005–1.0 ng mL^{-1}. Moreover, the fabricated aptasensor showed a high selectivity for oxytetracycline in the presence of similar drugs like tetracycline, doxycycline, and chlortetracycline [59].

Similarly, label-free detection of tetracycline (TET) was performed using two aptasensors made of carbon paste electrode (CPE) with oleic acid (OA) and a magnetic bar carbon paste electrode (MBCPE) with Fe_3O_4 magnetic nanoparticles and oleic acid (OA) following the modification of electrode surfaces using anti-TET. The LoD were 1.0×10^{-12} to 1.0×10^{-7} M and 3.0×10^{-13} M for the two aptasensors, respectively, and the sensors were applied to pharmaceutical formulations, serum samples, as well as food products (milk and honey) [60].

A sensor based on nanocomposites of mC with SnO_x and TiO_2 nanocrystals was used to determine tobramycin (TOB) in urine and serum samples selectively and in the presence of kanamycin, oxytetracycline, and doxycycline. The aptasensor showed an excellent sensitivity with a LoD of 0.01 nM [61].

Chloramphenicol was also determined in eye drop formulations using N-doped graphene nano-sheet-Au NP composite (Au/N-G). The LoD was 0.59 μM, and the sensor showed a selectivity in the presence of interferences like oxytetracycline, chlortetracycline, ascorbic acid, and metronidazole [62]. Other applications included sulphamethoxazole using molecularly imprinted polymers (MIPs) decorated with Fe_3O_4 magnetic nanoparticles (MNPs) on SPCE [63].

Immunosensors for 17β-estradiol composed of Au electrode nanoparticle-thiolated protein G-scaffold. This structure has facilitated the anchoring of a mouse monoclonal anti-estradiol antibody. The LoD was 26 pg mL^{-1}. As per the authors, square wave voltammetry (SWV) was more sensitive (18 pg mL^{-1}) and required less time and effort compared to EIS [64].

Target drug	Sensing interface	Electrode	Matrix	Sensing measurement method and strategy	LoD	Ref
Raloxifene	Nd_2O_5 NPs@GO/GCE	GCE	ND Serum and urine	EIS CV Amperometry (Label-free)	ND ND 18.43 nM	[58]
OTC	Aptasensor (Fe_3O_4@mC_{900})	GCE	Milk samples	EIS (Label-free)	0.027 pg mL^{-1}	[59]
TET	Aptasensor 1: CPE/OA/anti-TET Aptasensor 2: MBCPE/Fe_3O_4NPs/OA/anti-TET	CPE MBCPE	Tablets, milk, honey, and serum	EIS (Label-free)	10^{-1}–10^{-7} M 3.0×10^{-13} M	[60]
TOB	Aptasensor/ SnO_x@TiO_2@mC	GCE	Urine and serum	EIS (Label-free)	0.01 nM	[61]
Chloramphenicol	Au/N-G	GCE	Eye drops	EIS (Label-free)	0.59 μM	[62]
Sulphamethoxazole	MIP-decorated Fe_3O_4 MNPs	SPCE	Seawater	EIS (Label-free)	0.001 nM	[63]
17β-estradiol	Au nanoparticle-thiolated protein G-scaffold	Au	Serum	EIS (Label-free)	26 pg $mL{-}^{1}$	[64]
BPA	AuNPs/PB/CNTs-COOH/GCE	GCE	Water	EIS (Labelled detection)	0.045 pM	[65]
P4	ssDNA/Au	Au	Tap water	EIS (Labelled detection)	0.90 ng $mL{-}^{1}$	[66]

Table 1.
Applications of EIS in analysis and characterizations of different drug materials in variable matrices.

Bisphenol A (BPA), a xenoestrogen with an estrogen-mimicking effect and that is widely used as a precursor in plastics industry, has been determined using a labelled aptasensor made of gold nanoparticles (AuNPs), Prussian blue (PB), and functionalized carbon nanotubes (AuNPs/PB/CNTs-COOH).

Determination of progesterone (P4) in water and other clinical samples was performed using single-stranded ssDNA aptamers with high binding affinity to P4 [66].

5. EIS in food analysis

In addition to food contamination with antimicrobials and other drugs, bacteria and other pathogens like mycotoxins (secondary metabolites of microfungi) or chemicals such as pesticides are also other sources of food contamination. Food contamination can occur at any stage of food production, storage, or dissemination. Sicknesses caused by foodborne pathogens include symptoms such as diarrhea, nausea, vomiting, septicemia, meningitis, and even death [50, 53, 67, 68]. Pathogens include famous strains of bacteria such as different species of Salmonella (e.g., *S. enteritidis* and *S. typhimurium*), *Escherichia coli* (*E. coli*), and *Staphylococcus aureus* (*S. aureus*).

Table 2 shows examples of different bacterial strains that have been determined in food products using EIS-based aptamers.

A highly specific DNA—aptamer to *S. enteritidis* in pork products—was developed using gold NPs, i.e., modified SPCE (GNPs-SPCE). The developed aptasensor

Target	Sensing interface	Electrode	Matrix	Sensing method	LoD	Ref
			Bacteria (LoD is measured as/CFU mL−1)*			
S. enteritidis	GNPs@SPCE	SPCE	Poultry products	EIS	600	[69]
S. typhimurium	GNPs@SPCE	SPCE	Animal-based products	EIS	600	[70]
Salmonella	GO+AuNPs@GCE	GCE	Pork meat	EIS	3.0	[71]
			Mycotoxins			
OTA	Diazonium modified-SPCE	SPCEs	Cocoa beans	EIS	0.15 ng mL^{-1}	[72]
OTA	Thiolated DNA aptamer	Au	Food products	EIS	0.12–0.40 nM	[73]
AFB_1	Cys-PAMAM-modified electrode	Au	Peanuts and corn snacks	EIS	0.40 ± 0.03 nM	[74]
			Pesticides			
Acetamiprid	Ag-NG/GCE	GCE	Cucumber and tomatoes	EIS	0.033 pM	[75]

**Colony-forming unit (CFU) mL^{-1}.*

Table 2.
Applications of EIS in analysis of food and food products.

was selective towards *S. enteritidis* and showed a negative response towards mixture of other pathogens [69]. Similarly, the same electrode was used as a sensor for *S. typhimurium* [70]. The developed sensors were capable of differentiating between the targeted *Salmonella* species (S. *enteritidis* and S. *typhimurium*) and the other *Salmonella*.

Another *Salmonella* sensor was fabricated using a GO/Au NP-modified GCE. The sensor was applied for pork samples and achieved a LoD of 3.0 colony-forming unit (CFU mL^{-1}) in this case [71] compared to 600 CFU mL^{-1} using the GNPs@SPCE aptasensors [69, 70].

The mycotoxin ochratoxin (OTA) has been determined in a variety of samples, e.g., in cocoa beans, using EIS aptasensor developed using a diazonium-coupling reaction mechanism for the immobilization of anti-OTA-aptamer on screen-printed carbon electrode (SPCE) [72]. EIS was also applied for the determination of OTA using a thiolated DNA aptamer immobilized by chemisorption to the surface of Au electrode [73]. Other mycotoxins, e.g., Aflatoxin B_1 (AFB_1) were detected using layer coating of cystamine (Cys), poly (amido-amine) dendrimers of generation 4.0 (PAMAM G4) and DNA aptamers (on Au electrode) specific to AFB_1 [74].

Pesticides, e.g., acetamiprid, were determined in samples of vegetables (tomatoes and cucumber) using AgNP-modified nitrogen-doped graphene (AgNPs/NG). The designed aptasensor was sensitive, selective, and economical and did not require intricate labelling procedures [75].

6. EIS in water and wastewater analysis

Discharge of heavy metals (HMs) into the water bodies via industrial activities and other sources, e.g., mining, acid rain, agricultural waste, etc., denotes a worldwide challenge. As previously mentioned in this chapter, HMs and other emergent contaminants possess a significant influence on the environment and human health. The intensifying flux of HMs into aquatic environments and the properties of HMs (toxicity, degradation rates, accumulation, uptake, bioavailability, etc.) necessitate the presence of firm rules and action plans for monitoring, detoxification methodologies, and treatment technologies to keep their concentrations within the permitted levels [23–26, 76].

Table 3 shows examples for the applications of EIS in determination of water contaminants such as HMs, pesticides, drugs, and pharmaceuticals.

EIS has been applied for quantitative determination of HMs in water samples. In one of the investigations, a bi-enzymatic biosensor was constructed by immobilizing *Arthrospira platensis* cells (Spirulina) on gold interdigitated transducers. Consequently, phosphatase and esterase activities were inhibited by HMs and pesticides, respectively. This approach was used to determine Hg^{2+} and Cd^{2+} as well as parathion, paraoxon, and triazine pesticides, alone or in mixture with the HMs [77].

In another approach, a three-electrode sensor was printed on a polyethylene terephthalate film (PET) and was applied for impedimetric determination of Pb^{2+} and Cd^{2+} in water samples at nanomolar level [78]. An electrochemical DNA biosensor based on microspheres of cuprous oxide (Cu_2O) and nano-chitosan (NC) was used for Hg^{2+} detection in river water samples with a LoD of 0.15 nM [79].

Other contaminants like pesticides and herbicides as well as drugs and pharmaceuticals were also determined using EIS [63, 65, 66, 75, 77] (**Table 3**).

Target	Sensing interface	Matrix	LoD	Ref
		Heavy metals		
Hg^{2+}, Cd^{2+}	*Arthrospira platensis* cells (Spirulina)	Municipal wastewater	10^{-20} M and 10^{-20} M	[77]
Pb^{2+}, Cd^{2+}	PET-SPE	Water	1 nM for both metals	[78]
Hg^{2+}	Cu_2O@NCs	River water samples	0.15 nM	[79]
		Pesticides and herbicides		
Parathion-methyl	*Arthrospira platensis* cells (Spirulina)	Municipal wastewater	10^{-20} M	[77]
Paraoxon-methyl	*Arthrospira platensis* cells (Spirulina)	Municipal wastewater	10^{-18} M	[77]
Triazine	*Arthrospira platensis* cells (Spirulina)	Municipal wastewater	10^{-20} M	[77]
Acetamiprid	Ag-NG/GCE	Wastewater	0.033 pM	[75]
		Drugs and pharmaceuticals		
Sulphamethoxazole	MIP-decorated Fe_3O_4 MNPs@SPCE	Seawater	0.001 nM	[63]
BPA	AuNPs/PB/CNTs-COOH/GCE	Water	0.045 pM	[65]
P4	ssDNA/Au	Tap water	0.90 ng mL^{-1}	[66]

Table 3.
Applications of EIS in analysis of water.

7. Conclusions

The literature is rich in articles and reviews that investigate the applications of electrochemical impedance spectroscopy in detections of various contaminants such as heavy metals, drugs, and pharmaceuticals, as well as pesticides. The advantages that impedance spectroscopy introduces as an electrochemical technique are innumerable. High sensitivity, specificity, selectivity, no time or effort consumption, and being label-free are the major advantages reported in the majority of the surveyed literature. As the mentioned contaminants usually exist as traces in complicated matrices, impedance spectroscopy with the mentioned advantages was usually the electrochemical technique of choice for the detection of these contaminants in water, food, and drug matrices.

Conflict of interest

The authors declare no conflict of interest.

Author details

Marwa El-Azazy
Department of Chemistry and Earth Sciences, College of Arts and Sciences, Qatar University, Doha, Qatar

*Address all correspondence to: marwasaid@qu.edu.qa

References

[1] Macdonald DB. Reflections on the history of electrochemical impedance spectroscopy. Electrochimica Acta. 2006;**51**:1376-1388

[2] Barsoukov E, Macdonald JR. Impedance Spectroscopy: Theory, Experiment, and Applications. 3rd ed. Hoboken, NJ: Wiley; 2018

[3] Pejcic B, De Marco R. Impedance spectroscopy: Over 35 years of electrochemical sensor optimization. Electrochimica Acta. 2006;**51**:6217-6229

[4] Randviir EP, Banks CE. Electrochemical impedance spectroscopy: An overview of bioanalytical applications. Analytical Methods. 2013;**5**:1098-1115

[5] Rupp GM, Opitz AK, Nenning A, Limbeck A, Fleig J. Real-time impedance monitoring of oxygen reduction during surface modification of thin film cathodes. Nature Materials. 2017;**16**:640

[6] Keddam M, Nóvoa XR, Vivier V. The concept of floating electrode for contact-less electrochemical measurements: Application to reinforcing steel-bar corrosion in concrete. Corrosion Science. 2009;**51**:1795-1801

[7] Alexander CL, Orazem ME. Indirect electrochemical impedance spectroscopy for corrosion detection in external post-tensioned tendons: 1. Proof of concept. Corrosion Science. 2020;**164**:108331

[8] Song D, Wan H, Tu X, Li W. A better understanding of failure process of waterborne coating/metal interface evaluated by electrochemical impedance spectroscopy. Progress in Organic Coating. 2020;**142**:105558

[9] Pettit CM, Garland JE, Etukudo NR, Assiongbon KA, Emery SB, Roy D. Electrodeposition of indium on molybdenum studied with optical second harmonic generation and electrochemical impedance spectroscopy. Applied Surface Science. 2002;**202**:33-46

[10] Frateur I, Cattarin S, Musiani M, Tribollet B. Electrodissolution of Ti and p-Si in acidic fluoride media: formation ratio of oxide layers from electrochemical impedance spectroscopy. Journal of Electroanalytical Chemistry. 2000;**482**:202-210

[11] Soaresa C, Tenreiro Machadob JA, Lopesc AM, Vieiraa E, Delerue-Matosa C. Electrochemical impedance spectroscopy characterization of beverages. Food Chemistry. 2020;**302**:125345

[12] Lopes AM, Machado JT, Ramalho E, Silva V. Milk characterization using electrical impedance spectroscopy and fractional models. Food Analytical Methods. 2018;**11**:901-912

[13] Liu X, Duckworth PA, Wong DK. Square wave voltammetry versus electrochemical impedance spectroscopy as a rapid detection technique at electrochemical immunosensors. Biosensors and Bioelectronics. 2010;**25**:1467

[14] Mishra SK, Kumar D, Biradar AM, Rajesh. Electrochemical impedance spectroscopy characterization of mercaptopropionic acid capped ZnS nanocrystal based bioelectrode for the detection of the cardiac biomarker—myoglobin. Bioelectrochemistry. 2012;**88**:118-126

[15] Rajesh, Sharma V, Tanwar VK, Mishra SK, Biradar AM. Electrochemical impedance immunosensor for the detection of cardiac biomarker

Myogobin (Mb) in aqueous solution. Thin Solid Films. 2010;**519**:1167-1170

[16] Finšgar M, Xhanari K, Petovar B. Copper-film electrodes for Pb(II) trace analysis and a detailed electrochemical impedance spectroscopy study. Microchemical Journal. 2019;**147**: 863-871

[17] Fan L, Zhao G, Shi H, Liu M, Li Z. A highly selective electrochemical impedance spectroscopy-based aptasensor for sensitive detection of acetamiprid. Biosensors & Bioelectronics. 2013;**4315**:12-18

[18] Sacco A. Electrochemical impedance spectroscopy: Fundamentals and application in dye-sensitized solar cells. Renewable and Sustainable Energy Reviews. 2017;**79**:814-829

[19] Kumar S, Singh PK, Chilana GS. Study of silicon solar cell at different intensities of illumination and wavelengths using impedance spectroscopy. Solar Energy Materials & Solar Cells. 2009;**93**:1881-1884

[20] Halme J, Vahermaa P, Miettunen K, Lund P. Device physics of dye solar cells. Advanced Materials. 2010;**22**:E210-E234

[21] Scrosati B, Croce F, Persi L. Impedance spectroscopy study of PEO-based nanocomposite polymer electrolytes. Journal of the Electrochemical Society. 2000;**147**:1718-1721

[22] Vakili M, Rafatullah M, Salamatinia B, Abdullah AZ, Ibrahim MH, Tan KB, et al. Application of chitosan and its derivatives as adsorbents for dye removal from water and wastewater: A review. Carbohydrate Polymers. 2014;**113**:115-130

[23] Zhang L, Zeng Y, Cheng Z. Removal of heavy metal ions using chitosan and modified chitosan: A review. Journal of Molecular Liquids. 2016;**214**:175-191

[24] Tchounwou PB, Yedjou CG, Patlolla AK, Sutton DJ. Heavy metals toxicity and the environment. EXS. 2012;**101**:133-164

[25] Järup L. Hazards of heavy metal contamination. British Medical Bulletin. 2003;**68**:167-182

[26] U.S. EPA (Environmental Protection Agency). Mercury Study Report to Congress. 1997. Available from: http://www.epa.gov/mercury/report.htm

[27] El-Azazy M, Dimassi S, El-Shafie AS, Issa AA. Bio-waste *Aloe vera* leaves as an efficient adsorbent for titan yellow from wastewater: Structuring of a novel adsorbent using Plackett-Burman factorial design. Applied Sciences. 2019;**9**:1-20. DOI: 10.3390/app9224856

[28] El-Azazy M, El-Shafie AS, Ashraf A, Issa AA. Eco-structured biosorptive removal of basic Fuchsin using pistachio nutshells: A definitive screening design—Based approach. Applied Sciences. 2019;**9**:1-19. DOI: 10.3390/app9224855

[29] Singh KP, Singh AK, Singh UV, Verma P. Optimizing removal of ibuprofen from water by magnetic nanocomposite using Box-Behnken design. Environmental Science and Pollution Research. 2012;**19**:724-738. DOI: 10.1007/s11356-011-0611-4

[30] Aa N, Kommer GJ, Van Montfoort J, Versteegh J. Demographic projections of future pharmaceutical consumption in the Netherlands. Water Science and Technology. 2011;**63**:825-831. DOI: 10.2166/wst.2011.120

[31] Avcı A, Inci I, Baylan N. Adsorption of ciprofloxacin hydrochloride on multiwall carbon nanotube. Journal of Molecular Structure. 2020;**1206**:127711. DOI: 10.1016/j.molstruc.2020.127711

[32] Yu F, Sun S, Han S, Zheng J, Ma J. Adsorption removal of ciprofloxacin

by multi-walled carbon nanotubes with different oxygen contents from aqueous solutions. Chemical Engineering Journal. 2016;**285**:588-595. DOI: 10.1016/j.cej.2015.10.39

[33] Carabineiro S, Thavorn-Amornsri T, Pereira M, Serp P, Figueiredo J. Comparison between activated carbon, carbon xerogel and carbon nanotubes for the adsorption of the antibiotic ciprofloxacin. Catalysis Today. 2012;**186**:29-34. DOI: 10.1016/j.cattod.2011.08.020

[34] WHO. Worldwide strategy for the containment of antimicrobial resistance. WHO. 2001. Available from: http://whqlibdoc.who.int/hq/2001/WHO_CDS_CSR_DRS_2001.2a.pdf

[35] USDA. Foreign Agricultural Service. Veterinary drug MRL database. Available from: http://www.mrldatabase.com/default.cfm?selectvetdrug=1

[36] FAO/WHO. Available from: http://www.codexalimentarius.net/mrls/vetdrugs/jsp/vetd_q-s.jsp

[37] Lelieveld H, Motarjemi Y. Fundamentals in management of food safety in the industrial setting: Challenges and outlook of the 21st Century. Food Safety Management System. 2014:1-20. DOI: 10.1016/B978-0-12-381504-0.00001-9

[38] Commission Regulation 1353/2007/EC of 20th November, amending Council Regulation 2377/90/EC laying down a Community procedure for the establishment of maximum residue limits of veterinary medicinal products in foodstuffs of animal origin. Official Journal, L303

[39] Elazazy MS, Ganesh K, Sivakumar V, Huessein YHA. Interaction of p-synephrine with p-chloranil: experimental design and multiple response optimization. RSC Advances. 2016;**6**:64967-64976. DOI: 10.1039/C6RA10533E

[40] Aly H, El-Shafie AS, El-Azazy M. Utilization of 7-chloro-4-nitrobenzo-2-oxa-1, 3-diazole (NBD-Cl) for spectrochemical determination of l-ornithine: A multivariate optimization-assisted approach. RSC Advances;**9**(38):22106-22115

[41] Al-Saad K, El-Azazy M, Issa AA, Al-Yafei A, El-Shafie AS, Al-Sulaiti M, et al. Recycling of date pits into a green adsorbent for removal of heavy metals: A fractional factorial design-based approach. Frontiers in Chemistry. 2019;7:552. DOI: 10.3389/fchem.2019.00552

[42] El-Azazy M, Kalla RN, Issa AA, Al-Sulaiti M, El-Shafie AS, Shomar B, et al. Pomegranate peels as versatile adsorbents for water purification: Application of Box–Behnken design as a methodological optimization approach. Environmental Progress & Sustainable Energy. 2019;**38**:1-12. DOI: 10.1002/ep.13223

[43] Elazazy MS, Shalaby A, Elbolkiny MN, Khalil HM. Performance and characteristics of a new metoclopramide hydrochloride selective membrane electrode. Chinese Pharmaceutical Journal. 2004;**56**:49-56

[44] Elazazy MS, Shalaby A, Elbolkiny MN, Khalil HM. Development of high performance ion-selective electrode for in situ measurements. Chinese Pharmaceutical Journal. 2003;**55**:491-497

[45] Pu Q, Elazazy MS, Alvarez JC. Label-free detection of heparin, streptavidin, and other probes by pulsed streaming potentials in plastic microfluidic channels. Analytical Chemistry. 2008;**80**:6532-6536

[46] Elazazy MS, El-Mammli MY, Shalaby A, Ayad MM. Conductometric

determination of some important carboxylic acid derivatives and hydrochlorides in pharmaceutical formulations. Chemia Analityczna. 2008;**53**:725-736

[47] Jia W, Chu X, Ling Y, Huang J, Chang J. High-throughput screening of pesticide and veterinary drug residues in baby food by liquid chromatography coupled to quadrupole Orbitrap mass spectrometry. Journal of Chromatography. A. 2014;**1347**:122-128. DOI: 10.1016/j.chroma.2014.04.081

[48] Zhang B, Han X, Gu P, Fang S, Bai J. Response surface methodology approach for optimization of ciprofloxacin adsorption using activated carbon derived from the residue of desilicated rice husk. Journal of Molecular Liquids. 2017;**238**:316-325. DOI: 10.1016/j.molliq.2017.04.022

[49] He X, Wang GN, Yang K, Liu HZ, Wu XJ, Wang JP. Magnetic graphene dispersive solid phase extraction combining high performance liquid chromatography for determination of fluoroquinolones in foods. Food Chemistry. 2017;**221**:1226-1231. DOI: 10.1016/j.foodchem.2016.11.035

[50] Li F, Yu Z, Han X, Lai RY. Electrochemical aptamer-based sensors for food and water analysis: A review. Analytica Chimica Acta. 2019;**1051**:1-23

[51] Tóth K, Štulík K, Kutner W, Fehér Z, Lindner E. Electrochemical detection in liquid flow analytical techniques: Characterization and classification (IUPAC Technical Report). Pure and Applied Chemistry. 2004;**76**:1119-1138

[52] Su L, Jia WZ, Hou CJ, Lei Y. Microbial biosensors: a review. Biosensors & Bioelectronics. 2011;**26**:1788-1799

[53] Vasilescu A, Marty J. Electrochemical aptasensors for the assessment of food quality and safety. TrAC, Trends in Analytical Chemistry. 2016;**79**:60-70

[54] Ramnani P, Saucedo NM, Mulchandani A. Carbon nanomaterial-based electrochemical biosensors for label-free sensing of environmental pollutants. Chemosphere. 2016;**143**:85-89

[55] Ebele AJ, Abdallah MA, Harrad S. Pharmaceuticals and personal care products (PPCPs) in the freshwater aquatic environment. Emerging Contaminants. 2017;**3**:1-16. DOI: 10.1016/j.emcon.2016.12.004

[56] Boyer EW. Management of opioid analgesic overdose. The New England Journal of Medicine. 2012;**367**:146-155. DOI: 10.1056/NEJMra1202561

[57] Carbajo JB, Petre AL, Rosal R, Herrera S, Leton P, Garcia-Calvo E, et al. Continuous ozonation treatment of ofloxacin: Transformation products, water matrix effect and aquatic toxicity. Journal of Hazardous Materials. 2015;**292**:34-43. DOI: 10.1016/j.jhazmat.2015.02.075

[58] Chena T-W, Merlin JP, Chen S-M, Anandaraj S, Elshikh MS, Tseng T-W, et al. Sonochemical synthesis and fabrication of neodymium sesquioxide entrapped with graphene oxide based hierarchical nanocomposite for highly sensitive electrochemical sensor of anti-cancer (raloxifene) drug. Ultrasonics Sonochemistry. 2020;**64**:104717

[59] Song Y, Duan F, Zhang S, Tian J-Y, Zhang Z, Wang Z-W, et al. Iron oxide@ mesoporous carbon architectures derived from an Fe(ii)-based metal organic framework for highly sensitive oxytetracycline determination. Journal of Materials Chemistry A. 2017;**5**:19378-19389

[60] Jahanbani S, Benvidi A. Comparison of two fabricated aptasensors based on modified carbon paste/oleic acid and

magnetic bar carbon paste/Fe_3O_4@ oleic acid nanoparticle electrodes for tetracycline detection. Biosensors & Bioelectronics. 2016;**85**:553-562

[61] Wang M, Hu B, Yang C, Zhang Z, He L, Fang S, et al. Electrochemical biosensing based on protein-directed carbon nanospheres embedded with SnO_x and TiO_2 nanocrystals for sensitive detection of tobramycin. Biosensors & Bioelectronics. 2018;**99**:176-185

[62] Borowiec J, Wang R, Zhu L, Zhang J. Synthesis of nitrogen-doped graphene nanosheets decorated with gold nanoparticles as an improved sensor for electrochemical determination of chloramphenicol. Electrochimica Acta. 2013;**99**:138-144

[63] Zamora-Gálvez A, Ait-Lahcen A, Mercante LA, Morales-Narváez E, Amine A, Merkoçi A. Molecularly imprinted polymer-decorated magnetite nanoparticles for selective sulfonamide detection. Analytical Chemistry. 2016;**88**:3578-3584

[64] Liu X, Duckworth PA, Wong DK. Square wave voltammetry versus electrochemical impedance spectroscopy as a rapid detection technique at electrochemical immunosensors. Biosensors & Bioelectronics. 2010;**25**:1467

[65] Azadbakht A, Roushani M, Abbasi AR, Derikvand Z. A novel impedimetric aptasensor, based on functionalized carbon nanotubes and Prussian blue as labels. Analytical Biochemistry. 2016;**512**(2016):58-69

[66] Jiménez GC, Eissa S, Ng A, Alhadrami H, Zourob M. Aptamer-based label-free impedimetric biosensor for detection of progesterone. Analytical Chemistry. 2015;**87**:1075-1082

[67] Yeni F, Yavas S, Alpas H, Soyer Y. Most common foodborne pathogens and mycotoxins on fresh produce: A review of recent outbreaks. Critical Reviews in Food Science and Nutrition. 2016;**56**:1532-1544

[68] Sharma H, Mutharasan R. Review of biosensors for foodborne pathogens and toxins. Sensors and Actuators B: Chemical. 2013;**183**:535-549

[69] Labib M, Zamay AS, Kolovskaya OS, Reshetneva IT, Zamay GS, Kibbee RJ, et al. Aptamer-based impedimetric sensor for bacterial typing. Analytical Chemistry. 2012;**84**:8114-8117

[70] Labib M, Zamay AS, Kolovskaya OS, Reshetneva IT, Zamay GS, Kibbee RJ, et al. Aptamer-based viability impedimetric sensor for bacteria. Analytical Chemistry. 2012;**84**:8966-8969

[71] Ma XY, Jiang YH, Jia F, Yu Y, Wang ZP. An aptamer-based electrochemical biosensor for the detection of Salmonella. Journal of Microbiological Methods. 2014;**98**:94-98

[72] Mishra RK, Hayat A, Catanante G, Ocana C, Marty J-L. A label free aptasensor for ochratoxin a detection in cocoa beans: An application to chocolate industries. Analytica Chimica Acta. 2015;**889**:106-112

[73] Castillo G, Lamberti I, Mosiello L, Hianik T. Impedimetric DNA aptasensor for sensitive detection of ochratoxin a in food. Electroanalysis. 2012;**24**:512-520

[74] Castillo G, Spinella K, Poturnayová A, Snejdárková M, Mosiello L, Hianik T. Detection of aflatoxin B_1 by aptamer-based biosensor using PAMAM dendrimers as immobilization platform. Food Control. 2015;**52**:9-18

[75] Jiang D, Du XJ, Liu Q, Zhou L, Dai LM, Qian J, et al. Silver nanoparticles anchored on nitrogen-doped graphene as a novel electrochemical biosensing platform

with enhanced sensitivity for aptamer-based pesticide assay. The Analyst. 2015;**140**:6404-6411

[76] Fu F, Wang Q. Removal of heavy metal ions from wastewaters: A review. Journal of Environmental Management. 2011;**92**:407-418. DOI: 10.1016/j.jenvman.2010.11.011

[77] Tekaya N, Saiapina O, Ouada HB, Lagarde F, Namour P, Ouada HB, et al. Bi-enzymatic conductometric biosensor for detection of heavy metal ions and pesticides in water samples based on enzymatic inhibition in Arthrospira *platensis*. Journal of Environmental Protection. 2014;**5**:441-453

[78] Avuthu SGR, Narakathu BB, Eshkeiti A, Emamian S, Bazuin BJ, Joyce M, et al. Detection of heavy metals using fully printed three electrode electrochemical sensor. IEEE Sensors Journal. 2014;**10**:669-672

[79] Liu S, Kang M, Yan F, Peng D, Yang Y, He L, et al. Electrochemical DNA biosensor based on microspheres of cuprous oxide and nano-chitosan for Hg(II) Detection. Electrochimica Acta. 2015;**160**:64-73